# SMOG CHECK

## *Science, Federalism, and the Politics of Clean Air*

DOUGLAS S. EISINGER

Washington, DC • London

First published in 2010 by RFF Press, an imprint of Earthscan



Earthscan LLC, 1616 P Street, NW, Washington, DC 20036, USA
Earthscan Ltd, Dunstan House, 14a St Cross Street, London EC1N 8XA, UK
Earthscan publishes in association with the International Institute for Environment and Development

For more information on RFF Press and Earthscan publications, see www.rffpress.org and www.earthscan.co.uk or write to earthinfo@earthscan.co.uk

ISBN: 978-1-933115-71-9 (hardback)
ISBN: 978-1-933115-72-6 (paperback)

Copyedited by Joyce Bond
Typeset by OKS
Cover design by Ellen A. Davey and Clifford Hayes

Library of Congress Cataloging-in-Publication Data

Eisinger, Douglas S.
Smog check : science, federalism, and the politics of clean air / Douglas S. Eisinger.
p. cm.
Includes bibliographical references and index.
ISBN 978-1-933115-71-9 (hardback : alk. paper) – ISBN 978-1-933115-72-6 (pbk. : alk. paper)
1. Air–Pollution–Government policy–United States. 2. Automobiles–Motors–Exhaust gas–California.
3. Automobiles–Inspection–California. 4. Federal-state controversies–United States.
I. Title.
HC110.A4E46 2010
363.739'25610973–dc22 2010007293

A catalogue record for this book is available from the British Library

At Earthscan we strive to minimize our environmental impacts and carbon footprint through reducing waste, recycling and offsetting our $CO_2$ emissions, including those created through publication of this book. For more details of our environmental policy, see www.earthscan.co.uk.

Printed and bound in the UK by TJ International, an ISO 14001 accredited company.
The paper used is FSC certified and the inks are vegetable based.

The findings, interpretations, and conclusions offered in this publication are those of the authors. They do not necessarily represent the views of Resources for the Future, its directors, or its officers. Similarly, any geographic boundaries and titles depicted in RFF Press publications do not imply any judgment or opinion about the legal status of a territory on the part of Resources for the Future.

# *About* Resources for the Future *and* RFF Press

**Resources for the Future (RFF)** improves environmental and natural resource policymaking worldwide through independent social science research of the highest caliber. Founded in 1952, RFF pioneered the application of economics as a tool for developing more effective policy about the use and conservation of natural resources. Its scholars continue to employ social science methods to analyze critical issues concerning pollution control, energy policy, land and water use, hazardous waste, climate change, biodiversity, and the environmental challenges of developing countries.

**RFF Press** supports the mission of RFF by publishing book-length works that present a broad range of approaches to the study of natural resources and the environment. Its authors and editors include RFF staff, researchers from the larger academic and policy communities, and journalists. Audiences for publications by RFF Press include all of the participants in the policymaking process—scholars, the media, advocacy groups, NGOs, professionals in business and government, and the public. RFF Press is an imprint of **Earthscan**, a global publisher of books and journals about the environment and sustainable development.

# CONTENTS

# ACKNOWLEDGMENTS

This work was completed over several years with the generous support of many individuals. Don Reisman, director of Resources for the Future (RFF) Press, patiently provided writing advice; his insights substantially improved the book. Part of the work was supported while I was an RFF fellow in environmental regulatory implementation. The fellowship was overseen by Dr. Molly Macauley at RFF and was made possible by a grant from the Andrew W. Mellon Foundation.

I am indebted to Peter Wathern, professor emeritus at the University of Wales, Aberystwyth. Peter, far removed in place and time from the events discussed here, provided objective wisdom and much-needed humor and encouragement. Thanks also to my University of Hawaii colleagues Jackie Miller, Deborah Woodcock, and Peter Flachsbart. Jackie connected me with Peter Wathern; Deborah, who heard an early rendition of the story during the Shunzo Sakamaki Extraordinary Lecture Series, encouraged my efforts to set it down on paper. Peter Flachsbart helped fact-check emissions information.

Appreciation goes to my colleagues at the air quality research firm Sonoma Technology, Inc. (STI). Special thanks to Don Blumenthal and Lyle Chinkin for their book comments and to the STI support staff, Kristen Force, Wendy

Osmann, Jana Schwartz, Mary Anne Slocum, and Sandy Smethurst, as well as to Ken Craig, who helped fact-check the status of state inspection programs.

Heartfelt thanks to my former EPA colleagues Dave Howekamp, Felicia Marcus, Mary Nichols, and Dick Wilson, who gave generously of their time to relive the *Smog Check* experience and to contribute to this work. Numerous EPA Region 9 colleagues worked to support us at the time of the dispute. Unfortunately, many are not named in the text; peer reviewers counseled against including too many participants in a story that, if not stripped to its essentials, would be hard to follow. In partial recompense, I mention them here: Angela Baranco, Virginia Donahue, Sylvia Dugré, Bill Glenn, Roxanne Johnson, Bill Jones, Eleanor Kaplan, Sara Russell, Nina Spiegelman, Stephanie Valentine, and Bethany Whitaker.

Thanks also to Deb Niemeier at the University of California-Davis, John Harrison of the University of Hawaii's Environment Center, and Bob O'Loughlin at the U.S. Federal Highway Administration; each helped champion this project to RFF. Peer reviewers provided valuable feedback, among them Michael Hill, emeritus professor at the University of Newcastle; Dennis Pirages, professor at the University of Maryland; Walter A. Rosenbaum, emeritus professor at the University of Florida; and several anonymous reviewers. Birgit Wolff at Colorado State University funded my sharing the story at the Intercontinental Mobile Sources/Clean Air Conference in Madrid, Spain. Michael Kraft of the University of Wisconsin–Green Bay offered book proposal advice. Much appreciation to Joyce Bond for her manuscript edits and to Mike Oliver for his early editorial support.

Thanks to former California state assemblyman Richard Katz and former California state senator Quentin Kopp for reviewing the manuscript and contributing to this work. Staff to state legislators shaped the story's outcome. In particular, John Stevens, an aide to then assemblyman Richard Katz, and Carla Anderson, an aide to then state senator Robert Presley, deserve recognition for helping resolve the federal-state dispute discussed here.

Over the years, numerous colleagues responded graciously to information requests; special thanks to Steve Cadle of General Motors, Doug Lawson of the National Renewable Energy Laboratory, Jeff Long of the California Air Resources Board, Paul Roberts of STI, and Don Stedman of the University of Denver. Don Stedman also offered constructive comments on the book, as did Phil Lorang and Don Zinger of EPA.

Many thanks to Gail McTaggart and Lance Rogers; they provided inspiration as teachers and a lifelong set of warm memories. To my parents I owe everything. Finally, thanks go to my devoted wife, Patti Ai, and to my remarkable son, Andy, for being ever cheerful throughout this marathon. I hope we will look back and grin a bit whenever we get our car smog checked.

# LIST OF ACRONYMS AND ABBREVIATIONS

| | |
|---|---|
| AB | Assembly Bill |
| ASM | acceleration simulation mode |
| BAR | California Bureau of Automotive Repair |
| CAA | Clean Air Act |
| CAAA | Clean Air Act Amendments |
| CalEPA | California Environmental Protection Agency |
| CARB | California Air Resources Board |
| CO | carbon monoxide |
| $CO_2$ | carbon dioxide |
| EC | European Commission |
| EDF | Environmental Defense Fund |
| EPA | U.S. Environmental Protection Agency |
| E-REGS | Environmental Regulatory Spectrum |
| FIP | federal implementation plan |
| FTP | Federal Test Procedure |
| GAO | U.S. General Accounting Office (renamed Government Accountability Office) |
| GHG | greenhouse gas |
| GPC | gross polluter certification |
| HC | hydrocarbon(s) |
| HEP | high-emitter profile |

| | |
|---|---|
| I/M | inspection and maintenance |
| IM240 | I/M dynamometer test lasting 240 seconds |
| IMRC | California Inspection and Maintenance Review Committee |
| LEV | Low Emission Vehicle (Program) |
| MOA | memorandum of agreement |
| MOBILE | U.S. EPA's on-road mobile source emissions model |
| MY | model year |
| NAAQS | National Ambient Air Quality Standards |
| NAS | U.S. National Academy of Sciences |
| NO | nitric oxide |
| $NO_2$ | nitrogen dioxide |
| $NO_x$ | oxides of nitrogen |
| NPR | National Performance Review |
| NRC | U.S. National Research Council |
| NRDC | Natural Resources Defense Council |
| OBD | on-board diagnostic (equipment) |
| OIG | U.S. EPA Office of Inspector General |
| OMS | U.S. EPA Office of Mobile Sources (renamed Office of Transportation and Air Quality) |
| ORD | U.S. EPA Office of Research and Development |
| OTR | ozone transport region |
| PM | particulate matter |
| $PM_{2.5}$ | particulate matter with an aerodynamic diameter of 2.5 microns or less |
| RSD | remote sensing device |
| SB | Senate Bill |
| SCAQMD | South Coast Air Quality Management District |
| SFTP | Supplemental Federal Test Procedure |
| SIP | state implementation plan |
| VOC | volatile organic compound |
| ZEV | Zero Emission Vehicle (Program) |

# FOREWORD

*Smog Check* is a work of public policy scholarship that reads at times like a gripping political thriller and other times like a bittersweet tale of love's labor lost. How can the story of one of the least-debated sections of the 1990 Clean Air Act Amendments, and the U.S. Environmental Protection Agency's efforts to implement it, hold so much drama?

Nothing about the nation's struggle for clean air is free of technical or political controversy, but the problem of pollution created by cars, from the moment they are driven off the lot until the time they are recycled as scrap, engages the ordinary person in ways that power plant scrubbers or new-car emissions standards do not. Accidental or deliberate tampering, poor maintenance or personal driving habits, even variability in the manufacturing process can cause a vehicle to fail an inspection and lead to expensive repairs that may or may not lead to lasting improvements in emissions. Whether administered by government-run or licensed private facilities, inspection stations will occasionally pass or fail a vehicle in error, and repair stations may overcharge for their work or fail to correct the problems that caused excess emissions. And any program that imposes requirements on all vehicle owners attracts scrutiny from radio talk-show hosts, legislators, and the driving public, all of whom are ready to assert that they understand the problem better than a bunch of bureaucrats.

Working for EPA, Doug Eisinger was one of those bureaucrats. Not only was he on point to get the Smog Check program in place in California just as a new president with strong ties to the state was coming into office, but he also brought an unusual combination of analytical skills and a historian's eye to the task. As he describes, what might seem an arcane debate about how cars should be inspected was actually the first round in a massive battle between determined bureaucrats and unmovable politicians, with the fate of the brand new Clean Air Act Amendments in the balance.

As one of the political appointees who played a big role in the ensuing conflict, I can attest to the level of political attention the conflict attracted—attention motivated by the resistance of California and other states to EPA's prescribed program. Resistance to EPA regulations led to the reinvention of Smog Check before EPA's required program could even begin.

Eisinger's detailed, factual account brings back memories I had successfully repressed. But it is his thoughtful assessment of the flaws in the legislative and regulatory processes that makes this a worthwhile book for any student or practitioner of public policy. In particular, his review of the controversy over the scientific data on which EPA engineers based their strongly held views has applications that go beyond the specific limitations of any one agency like EPA. And his observations about the gulf between congressional desire for micro-management and states' desire to independently administer federally mandated programs are not unique to the Clean Air Act.

The book also contains useful advice for those who would try to understand the relationships between career government regulators and their temporary political overseers. Although the Smog Check implementation battle played out in what was generally an EPA-friendly administration, the desire of a new Democratic White House, in trying to please its California base, contributed to instances of mishandled communications and information sharing.

As an environmental lawyer who has now spent more years inside government than out, I tend to avoid accounts of environmental policy successes or failures by academics, because they are too theoretical to be useful for implementation or too lacking in real-world understanding. This book, however, is both honest and useful. *Smog Check* is a rare and valuable glimpse at how America makes—or fails to make—critical environmental and public health policy decisions. It is a true cautionary tale.

Mary Nichols
Chairman, California Air Resources Board

*During the 1993–1994 Smog Check debate, Mary Nichols served as President Bill Clinton's appointed assistant administrator for air and radiation for the U.S. Environmental Protection Agency.*

# PREFACE

Smog Check remains one of the most profound experiences of my career. During an intense 15-month window, while I was a section chief in the San Francisco office of the U.S. Environmental Protection Agency (EPA), I participated in an agonizing, perplexing, and at times undeniably thrilling professional adventure. The adventure involved resolving a heated debate between the state of California and EPA over how to improve California's Smog Check program to inspect motor vehicles and reduce their contribution to urban air pollution.

At the time of the debate, which took place in the early 1990s, cars were the single most important air pollution source, and EPA thought that improving programs like Smog Check was the single best action states could take to improve air quality. The conflict flared over how to make those improvements. Since that time, I have had numerous opportunities to lecture about Smog Check, and the case has proven to have a timeless attraction. One of the reasons the Smog Check story holds such appeal is that it allows others to experience a tumultuous confrontation—a pitched policy battle between the state and federal governments. It also remains fascinating because it is such an all-encompassing case study of the policy process. The story involves the legislative process, regulatory development, policymaking negotiations, interactions with the press, relationships among career and appointed staff, scientific research, policy implementation and public reactions, and long-term policy evolution.

My challenge in writing this book was to meaningfully and objectively share and interpret the Smog Check story. Beyond documenting the remarkable Smog Check conflict itself, I also wanted to tell readers what happened afterward—the long-term outcomes from the intense debate that engulfed EPA and California and, with them, the rest of the nation. Finally, to make sense of the whole experience, for myself and others, I needed to draw lessons from Smog Check that would have lasting value. Those goals shaped the work here.

It is my hope that the story and its interpretation contribute to the public policy and air quality management fields. On a personal level, I also hope that those who were embroiled in Smog Check so many years ago find that through this book, that challenging time has borne insights that make those long-ago efforts more worthwhile.

Doug Eisinger
San Anselmo, California

"Odyssey: a long wandering or voyage usually marked by many changes of fortune."

*Merriam-Webster's Online Dictionary, 2010*

"If it's one thing I've learned about California politics, it's never mess with people's guns or their cars."

*California governor Jerry Brown, 1982*

The art of Smog Check (originally appeared in the *Los Angeles Times*).

# PART I
# BACKGROUND

CHAPTER 1

# INTRODUCTION TO A STATE VS FEDERAL DISPUTE

Smog Check was a policymaking odyssey. The story involves elected, appointed, and career government staff working under tremendous legal and political pressure. Through challenging times, they crafted and implemented a politically unpopular environmental program, then reshaped its design through the years to improve public acceptance and performance. They also made mistakes, which should come as no surprise, because mistakes are inevitable in any complicated endeavor. What makes Smog Check fascinating is the way the policy process unfolded, and why. A federal regulation required states to reduce pollution from cars and spelled out, in exhaustive detail, how that goal should be accomplished. Most states reluctantly accepted the federal demand; one in particular did not. The result was a highly visible and damaging confrontation between the state of California and the U.S. Environmental Protection Agency (EPA). The confrontation involved political brinksmanship, scientific intrigue, and even natural disaster. The fallout from the conflict and its negotiated resolution lasted many years, affected programs throughout the United States, and nearly upended the Clean Air Act – the U.S. statutory framework to address air pollution.

The debate involved motor vehicle inspection and maintenance (I/M). An I/M program measures a car's air pollutants and, if emissions exceed allowable levels, encourages vehicle repairs. Smog Check is the name of the I/M program in California. Many states have long had such programs. The requirement to implement more sophisticated I/M, or as Congress called it, an "enhanced" program, was among the most important provisions in the 1990 Clean Air Act Amendments (CAAA), which, at the time of the debate, had recently been approved. California's and EPA's divergent views on I/M led to the debate that is the focus of this book. EPA promoted the use of automotive test centers that separated inspections from repairs; California wanted small, garage-based businesses to be able to test *and* repair vehicles at the same location. This book recounts the Smog Check debate and its negotiated resolution, traces the real-world outcomes that resulted following the negotiations, and examines how the setting of environmental policies can be improved. Although the story involves air pollution control and motor vehicles, Smog Check's lessons apply across a wide spectrum of policy issues, especially those that require federal and state collaboration. Essentially, the narrative is a primer on policy.

During the debate, I managed the EPA field team responsible for I/M policy implementation in the southwestern United States. It was our job to inform California's leaders about federal requirements and help fulfill EPA's responsibility to enforce federal law.

## THE RELEVANCE OF THE SMOG CHECK DISPUTE TODAY

The Smog Check debate, and the negotiations that resolved it, took place in 1993 and 1994, shortly after Congress passed major changes to the Clean Air Act (in 1990) and EPA published enhanced I/M regulations (in 1992) that set ambitious goals for vehicle inspection programs. During the dispute, debate participants made numerous claims and counterclaims about the expected impacts of their preferred policies. With the benefit of hindsight, this study contrasts what policymakers predicted against what actually occurred following the conflict's resolution. In addition to recounting policy outcomes for California's Smog Check program, the book also examines the resulting national I/M policy experience and assesses lessons learned.

The lessons from Smog Check and its outcomes remain relevant, years after the California-EPA debate, for many environmental reasons. Notably, cars overwhelmingly contribute to air pollution problems around the world, and I/M programs are one of the few tools available to reduce pollution from vehicles already on the road. Smog Check's air pollution lessons, however, are dwarfed by its policy lessons. The most compelling reason Smog Check remains relevant over time is that the story involves fundamental questions concerning policy analysis, development, and implementation, including the following:

- how authority and power should be apportioned between state and federal governments;
- how government agencies should make decisions in the face of scientific uncertainty;
- when agencies need to negotiate and compromise to address differing views;
- what tools can be used to bridge deep-rooted policy disagreements; and
- what the probability is that forecast policy outcomes will materialize as predicted.

Insights from the Smog Check story and its outcomes are especially relevant given the challenge of addressing global climate change. Climate change is an air pollution problem. Through the burning of fossil fuels, as well as other human activities, greenhouse gases (GHGs)—particularly carbon dioxide ($CO_2$)—are emitted into the atmosphere and alter the climate on a global scale. The complexity of the climate change problem motivates governments to identify and employ all of the policy and technical lessons available from past air pollution control experiences. Smog Check and the overall U.S. I/M policy experience provide an enlightening case study that involves one of the core air pollution control programs contained in the U.S. Clean Air Act.

The case study's central theme involves policy structure. In some forums, political leaders debate whether to support regulation or deregulation. This work is built on the premise that environmental problems demand government attention, and that it is more important to design environmental regulations well than to debate whether to regulate at all. Through the Smog Check debate and its national ramifications, this study explores the question, when should the federal government mandate specific state actions, and when should it allow states discretion, provided state actions yield desired results? As discussed in Chapter 10, this work introduces a new tool called E-REGS, for the Environmental Regulatory Spectrum. E-REGS helps policy analysts and practitioners weigh where to place regulatory efforts along a spectrum that ranges from narrowly prescribed requirements to performance-measure-based mandates.

It is reasonable to ask whether this is a book about a policy success or failure. The answer to that question needs some context. The U.S. air quality management record is one of substantial success. Yet air quality control programs have not achieved all of their forecast benefits. As scientific and policy understanding evolves, air quality managers often learn that pollution sources emit more than previously thought, and control programs prove less effective than hoped.

As a result, U.S. air quality management can be characterized as establishing policy goals and deadlines; designing and implementing programs; making steady progress toward clean air; and then revising goals, deadlines, and programs to reflect new insights and respond to implementation difficulties. In this context, successful programs are those that consistently contribute toward goals, achieve much of what they were intended to accomplish, and cost—in money, political acceptability, and time—reasonably close to what was originally envisioned.

Even set against this fuzzy interpretation of success, however, the California Smog Check program, and the U.S. I/M experience as a whole, is more a policy failure than a success when weighed against the program goals EPA set in 1992. The programs described here fell short (half or less) of their hoped-for benefits, took years longer to implement, and incurred far greater costs—especially political costs—than envisioned. Many years after the conflict, I/M programs provided important ongoing emission reductions that improved urban-area air quality, but at a smaller scale than originally desired. In the book's epilogue, which includes viewpoints from others who served at EPA, former career official Dick Wilson admits, "I think that we probably made a mistake to spend so much time and political energy (and chips) on enhanced I/M," and former political appointee Felicia Marcus notes, "Enhancing I/M was a no-brainer and the right thing to do as a policy call. How we sought to implement it was the problem." If the clock could be turned back, EPA would not do the same things again.

## SMOG, THE 1990 CLEAN AIR ACT AMENDMENTS, AND I/M

Urban areas have long been troubled by air pollution problems. Beginning in the latter half of the twentieth century, urban air pollution was frequently referred to as *smog*, a word formed by combining "smoke" and "fog." Although there are a wide range of air quality problems that vary by location, time of day, season, and other factors, broadly speaking, urban-area smog usually refers to ozone, although it may also include some combination of acid vapor, air toxics, carbon monoxide (CO), nitrogen dioxide ($NO_2$), particulate matter (PM), or other compounds that impair human health and result from automotive and other emissions. Some of these pollutants, such as ozone and CO, are invisible gases; when individuals see a brownish haze obscuring their vision on a smoggy day, they are primarily viewing the impact of airborne $NO_2$ and fine liquid or solid particles—PM—that scatter and absorb visible light. Because the same sources and meteorological conditions contribute to many air pollution problems, a smoggy day can have high levels of multiple pollutants. Important human-caused air pollution sources include motor vehicles, industrial operations, and a wide array of smaller-scale business and personal activities. Natural sources, such as wind-blown dust, can also contribute.

U.S. efforts to achieve clean air have evolved over many decades and statutory attempts, and Chapter 2 provides a brief history of these efforts prior to the 1993–1994 Smog Check debate. Especially important to this story is the action taken by U.S. president George H.W. Bush on November 15, 1990, when he signed the 1990 CAAA into law. The 1990 CAAA established an ambitious agenda to achieve clean air throughout the United States within 20 years. The 1990 law triggered controls on virtually all pollution sources, from oil refineries to electric power plants to lawn mowers and the gasoline cans used to fill them. In particular, the new law focused on cars.

In 1990, cars were recognized as being one of the leading contributors to U.S. urban air pollution, and Congress sought to limit emissions from the vast fleet of vehicles already in use. Among its many requirements, the 1990 CAAA mandated enhanced motor vehicle I/M programs in heavily polluted areas.[1] Some of these areas, California included, already had I/M programs operating as of 1990; others did not. Congress tasked EPA with providing guidance to the states on how to implement a new, substantially improved version of I/M. EPA responded by requiring motorists to take their vehicles to an inspection center, where each vehicle could be tested and certified as meeting pollution standards. EPA said that inspection centers could do only testing; owners of vehicles that failed inspection would have to seek repairs separately, return to a test center for another inspection, and hope to receive certification the second time. EPA stated that its enhanced I/M program would "provide the largest emission reduction of any pollution control strategy EPA has thus far identified" (U.S. EPA 1992b). California opposed the EPA plan, sparking controversy and the debate described later in this book.

## CALIFORNIA'S UNIQUE ROLE IN AIR POLLUTION CONTROL

Before launching into a discussion of the debate, it is important to acknowledge that California is credited as an early and continuing air quality management pioneer. The U.S. National Research Council (NRC) found that since the early 1960s, California led the United States in requiring automotive emission control (Holmes et al. 2007). One researcher noted that at the outset of U.S. efforts to curb pollution, "no state was more instrumental to the formulation of national air pollution policy than California," and when Congress enacted a federal law to improve air quality, it protected California's regulatory autonomy so that the state could continue to act as a "laboratory for innovation" (Giovinazzo 2003, *900–901*). A publication on European environmental policymaking discussed how, beginning in the 1970s, "the USA became the benchmark for car emission regulations, with Europe and Japan initially lagging behind," and found that within the United States, California had a "forerunner status" with respect to automotive emission controls (Wurzel 2002, *93*). One article noted that California has the single largest U.S. automotive market and has led the nation in regulating automobiles: "In short, change the law in California and you can tip the entire national market" (Hertsgaard 2002, *7*).

In the 2000s, California expanded its leadership role to encompass climate change. The California Air Resources Board (CARB) established the first U.S. GHG automotive emission standards,[2] and the state enacted a landmark law to address climate change: Assembly Bill (AB) 32, the California Global Warming Solutions Act of 2006. When California governor Arnold Schwarzenegger signed AB 32 into law, the signing ceremony reflected California's international leadership in setting air quality benchmarks. Great Britain's prime minister, Tony Blair, observed the ceremony via a satellite connection; Japanese consul general Kazuo Kodama attended to represent Japan's prime minister, Junichiro Koizumi; and the premier of Manitoba, Canada, Gary Doer, spoke during the ceremony about how California's action created momentum throughout the world to address climate change problems.

Historically, California's leadership role has been driven by necessity: the state faces the most severe air pollution problems in the United States. Although the Smog Check program, as described here, has long encountered challenges, readers should appreciate that California's air pollution control efforts are generally recognized as preeminent.

## AIR POLLUTION AND MOTOR VEHICLES: AN ONGOING GLOBAL CHALLENGE

Despite efforts by California, the United States, and other nations, air pollution remains a policy challenge. Because this case study focuses on I/M, brief information about air pollution and automobiles will help readers appreciate where Smog Check and other I/M programs fit in the overall air quality management picture. This discussion highlights that the air quality problems that motivated enactment of the 1990 CAAA, and pollution from motor vehicles in particular, remain an ongoing concern in the United States and other nations.

### The Importance of Air Pollution

Despite years of progress in reducing emissions, at the start of the twenty-first century, at least half of the U.S. population lived in areas that failed to meet air quality standards. In 2009, for example, more than 126 million people lived in areas of the country that failed to meet ground-level (tropospheric) ozone standards. Ozone exposure is linked to respiratory problems and increased risk of death. Two pollutants—oxides of nitrogen ($NO_x$) and hydrocarbons (HC)—form ozone in the presence of sunlight; cars, trucks, and other sources emit $NO_x$ and HC.[3] Also in 2009, nearly 90 million people lived in U.S. areas failing to meet standards for very fine airborne particles, called $PM_{2.5}$. ($PM_{2.5}$ refers to particulate matter smaller than 2.5 microns in diameter; by comparison, a human hair is about 70 microns in diameter.) In addition to causing respiratory problems, fine particle exposure is associated with premature death among the elderly and reduced lung development in children. $PM_{2.5}$ is emitted directly by vehicles and other sources and is also formed in the atmosphere from pollutants such as $NO_x$. Another ongoing concern in the United States is exposure to cancer-causing air pollutants known as air toxics. In 2004, the NRC identified air toxics control, climate change, ozone, PM, regional haze, and air pollution problems unique to lower socioeconomic communities as among the most difficult air quality management challenges to be met in the decades ahead (Gauderman et al. 2004; NRC 2004; Jerrett et al. 2009; U.S. EPA 2009a, 2009c).

Other parts of the world experience air pollution problems even worse than those in the United States. For example, an investigation of 14 Chinese cities found "much higher levels of air pollution than [in] cities in developed countries" (Wang 2006, *4532*), and other researchers have noted that "air pollution is a serious public health problem in most major metropolitan areas in the developing world" (Faiz and Sturm 2000, *4745*).

## The Importance of Motor Vehicles

Although pollution problems vary across the developing and developed world, research consistently points to the importance of on-road motor vehicles. In the United States, for example, EPA estimated that as of 2008, on-road vehicles contributed approximately 25 to 50 percent of nationwide HC, $NO_x$, and CO emissions (U.S. EPA 2009b). In Southern California, home to the United States' most severe urban-scale air pollution, regulators estimated in 2008 that approximately 94 percent of the cancer risk associated with air toxics came from on- and off-road motor vehicles (SCAQMD 2008). In addition, research has established a link between proximity to major roadways and adverse health impacts for children and other people (e.g., Gauderman et al. 2007). Perhaps the most challenging environmental issue is concern over global climate change; transportation activities produced 30 percent of U.S. $CO_2$ emissions in 2008—one of the largest source categories in the country (U.S. EPA 2010).

Nor is the contribution of motor vehicles to air pollution unique to the United States. In their book *Two Billion Cars* (2009), researchers Daniel Sperling and Deborah Gordon noted that 16 of the world's 20 most polluted cities were in China, and that cars were responsible for more than half of the air pollution in Beijing and other affluent Chinese cities.

A prime concern is the worrisome trend of what some have characterized as the phenomenal growth in the number of motor vehicles, especially in the developing world (Faiz and Sturm 2000). For example, in a 2004–2005 report, the Indian government documented that one of the major sources of pollutants such as PM and nitrogen dioxide in Indian cities was vehicles, and that their number was "increasing exponentially" (Ministry of Environment and Forests 2005). India is but one example of a growing problem. As the authors of a World Bank study documented, "It is expected that the relative importance of mobile source air pollution will increase in developing countries as incomes grow" (Gwilliam et al. 2004, *3*).

## The Importance of I/M

Part of the challenge regarding motor vehicles is simply mathematics: even though automakers continue to build cleaner-operating cars, each year individuals throughout the world buy more vehicles and drive them at an ever-increasing rate. In the United States, the NRC found that "growth in vehicle miles traveled, personal automobile usage, and popularity of fuel-inefficient vehicles ... has offset a significant portion of the gains obtained from stricter emission standards on individual vehicles" (2004, *173*). As a consequence, regulators have developed a tool kit with half a dozen approaches to reducing automotive pollution, including new-vehicle emission standards;[4] diesel and gasoline fuel content requirements; travel demand management measures to encourage transit use, ridesharing, and other alternatives to driving; traffic flow improvements (because cars emit more in congestion); programs to retrofit or scrap older vehicles; and I/M.

I/M programs, the focal point for this book, attempt to identify highly polluting cars and trucks and require them to be repaired or replaced. Program effectiveness depends on the competency and honesty of inspectors and repair technicians, motorist compliance, inspection equipment, test stringency (pass–fail thresholds), and other factors (NRC 2001; Eisinger 2005). If implemented effectively, I/M programs are recognized as among the most cost-effective emission reduction options. Implementation problems have limited I/M program effectiveness, however, and scientists have long disputed program benefits, shaping the debate discussed in this book (Lawson 1993, 1995; NRC 2001; TRB 2002).

A special concern is the small percentage of highly polluting vehicles that contribute the vast majority of fleet emissions. Despite the implementation of increasingly stringent emission standards, research has consistently shown that 5 to 10 percent of vehicles contribute at least half of the emissions for any one pollutant such as HC, CO, and $NO_x$ (e.g., Stedman et al. 1991; Lawson et al. 1996; NRC 2001).

Different vehicles can contribute high emissions for different pollutants; therefore, it is difficult to identify or predict a single subset of vehicles requiring repairs. Thus, fleetwide vehicle inspections, tied to effective maintenance and repair, remain a key emission reduction strategy. In the United States, I/M programs are the main tools available to reduce pollution from the on-road vehicle fleet. As the NRC noted, however, "The existence of high emitters is a major challenge, and I/M programs have been less effective than expected in identifying high emitters" (2004, *172–173*). This study describes how the California-EPA debate sparked an evolution in U.S. I/M programs—an evolution that contributed directly to the NRC's findings that I/M has been less effective than expected.

## KEY ISSUES IN THE SMOG CHECK CONFLICT

Nine key issues shaped the California-EPA Smog Check debate. Some of these issues are scientific or technical in nature; others focus more on political and legal considerations, including the dividing line between federal and state authority. They are described here to give readers a better understanding of the conflict described in later chapters.

### The 50 Percent Discount: Testing and Repairs at Garages vs Centralized Testing

The most contentious issue of the debate was the so-called 50 percent discount: EPA's assertion that garage-based I/M programs were only half as effective at reducing emissions as centralized test systems. Centralized, or test-only, programs tested cars at centers run by either the government or private contractors. Vehicles that failed the test had to go elsewhere for repairs, and then return to the center for a retest. In contrast, garage-based, or test-and-repair, programs allowed state-authorized facilities to do both inspections and repairs. In 1992, of the 23 highly polluted states required to implement enhanced I/M, 12 states, California among them, had preexisting garage-based programs.

In its 1992 enhanced I/M regulations, EPA credited test-and-repair programs with only 50 percent of the emission reduction benefits of a test-only program. For example, EPA estimated a reduction in HC and CO emissions from cars of about 30 percent from test-only programs, but only about 15 percent from test-and-repair programs. Under EPA rules, highly polluted states that departed from a test-only design could not receive the emission reduction credits necessary to meet the 1990 CAAA. EPA's automotive specialists in Ann Arbor, Michigan, said the discount originated from EPA computer model algorithms. In effect, though, the 50 percent discount originated from the career staff's experience observing what had occurred across the United States at existing I/M programs, including the California garage-based Smog Check system.

California officials acknowledged that Smog Check experienced enforcement problems and test methods needed to be updated; however, the state's preference was to improve, rather than abandon, its existing program. At the outset of the debate, state regulators and legislators promoted a hybrid plan that would, if adopted, send vehicles to test-only stations for the first test, but then allow most failing vehicles to receive repairs *and retests* at state-sanctioned highly qualified facilities (garages) they planned to call Gold Shield stations. The Gold Shield plan directly contradicted EPA's goal to promote test-only I/M programs and was federally unacceptable given the 50 percent discount. The discount sparked vehement opposition to EPA's policy, as illustrated by correspondence from then California state senator Quentin Kopp to EPA management (Kopp 1993):

> I'm familiar with the undue efforts of the United States Environmental Protection Agency to impose absurd requirements upon California vehicle owners relative to the so-called Vehicle Inspection and Maintenance Program. I'm also aware of the EPA's arbitrary and capricious rule that any data emanating from a so-called decentralized system of inspection and maintenance must be discounted by 50 percent in terms of air quality results. Will you kindly explain the basis for such arbitrary and capricious regulation?

## Inspection Methods: EPA's IM240 Test vs California's ASM Test

In addition to favoring test-only programs, EPA preferred the use of specific test equipment and methods. Under the EPA requirements, vehicles were to be tested using a dynamometer. This device resembles a set of rolling pins placed side by side. A vehicle is hoisted onto the dynamometer, on which it can be driven while remaining motionless (see Figure 1-1).

EPA mandated that a vehicle be driven for four minutes on a dynamometer, while following an EPA-developed simulated trip called IM240 for the 240

**Figure 1-1. Front-wheel-drive Vehicle Receiving a Dynamometer-based I/M Test**
*Source:* Author photo; courtesy of A Smog Test Only Inc., Petaluma, California

seconds it took to complete the test. The IM240 test attempted to mimic real-world driving and emissions. The dynamometer-IM240 test package was expensive, about $145,000 in 1992; however, in a centralized test-only system, equipment costs could be amortized over many vehicles, minimizing individual inspection costs. From California's perspective, the key problem with these mandates was the expense of the IM240 test, which was prohibitive for the relatively low-volume private garages that wanted to participate in the program.

California sought to require the use of a less expensive dynamometer-based test, which it called the acceleration simulation mode (ASM) test. The state estimated that, at approximately $25,000 per unit in the early 1990s, ASM equipment was within the financial reach of private garages and could form the backbone of a test-and-repair-based Gold Shield program.[5] California believed the less expensive ASM test was nearly as effective as an IM240 test, a claim EPA disputed (CA IMRC 1993).

## Precedent: California vs Other States

EPA feared allowing California the freedom to establish a Gold Shield program, as it would set a precedent for the other 22 enhanced I/M states. Given California's preeminence in the air quality arena, if California established a hybrid program, other states would likely follow, thus eroding EPA's ability to promote test-only programs.

## Deadlines: States Were Racing the Clock to Implement I/M

The Smog Check debate took place in an atmosphere of urgency created by 1990 CAAA deadlines. Sacramento was one of several California areas required to implement enhanced I/M and is used here as an example to illustrate these deadlines. Under the 1990 CAAA, Sacramento had until November 1999 to demonstrate that it had met, or attained, the ozone standard. To demonstrate attainment, an area needed three consecutive years of acceptable air quality; this helped ensure that any one year's clean air was not due to unusual weather. Ozone is a summertime problem, and Sacramento had to achieve clean air by the start of summer in 1997 to meet the test of three clean-air years by November 1999. Seriously polluted areas like Sacramento depended on enhanced I/M to reduce emissions. Smog Check tested cars biennially, and thus it took a minimum of two years for the program to have an impact on the entire vehicle fleet. An enhanced Smog Check program therefore needed to begin in 1995 to contribute substantially to 1997 air quality improvements. Taking these deadlines into consideration, EPA mandated that by January 1, 1995, states needed to be able to test at least part of their vehicle fleet using enhanced I/M. EPA gave states until January 1, 1996, to make enhanced I/M fully operational.

In the early 1990s, Smog Check affected about 18 million California automobiles. Program changes took time to design and implement. States like California needed to settle on their program plans about a year in advance of the 1995 implementation deadline. In addition, Smog Check and other state I/M programs required authorizing legislation. To begin enhanced I/M planning in 1994, states ideally needed to pass legislation in 1993. Any early slip in the schedule, such as missing the legislative window to authorize enhanced IM, would compress the available window to design and implement a program and risk delaying attainment of the air quality standards.

## Remote Sensing Devices: New Technology Held the Promise of a Silver Bullet

Just prior to the Smog Check debate, two University of Denver scientists, Don Stedman and Gary Bishop, invented a new vehicle inspection technology called a remote sensing device (RSD) and tested its use in several locations, including Los Angeles (Lawson et al. 1990; Stedman et al. 1991). Stedman and Bishop set up their RSD units at the roadside. Much as a police officer uses a radar gun to measure a passing vehicle's speed, an RSD unit could measure a passing vehicle's emissions. RSD advocates portrayed the units as a potential silver bullet that could identify high-emitting vehicles while they were being driven. At the outset of the Smog Check debate, RSD technology was new, and many questions about its use and limitations were yet unanswered. Key California legislators hoped RSD would prove to be an unobtrusive and inexpensive method of finding polluting vehicles and perhaps even eliminate the need for traditional Smog Check inspections. EPA opposed giving substantial credit to what it viewed as an unproven and limited technology. EPA's view of RSD, pitted against California's interest in pioneering its use, created friction throughout the debate.

## Lack of Data: Little Technical Consensus Existed on the Best Way to Improve I/M

Despite widespread agreement in the early 1990s that California's existing I/M program was insufficient, there was little technical consensus on what was needed to improve Smog Check. EPA contended that its program was superior, but as the debate unfolded, it became clear the federal government lacked sufficient real-world data to substantiate its claims. California promoted less expensive test methods and innovative technology, but it too lacked the real-world data to prove their merits. Given the lack of a technical consensus on how to improve Smog Check, one logical outcome was to consider running pilot programs that could test the relative effectiveness of competing policy options. An important question that emerged was whether the state could adequately implement pilot studies and act on their results in time to meet implementation deadlines.

## States' Rights: Philosophical Differences between California and EPA

Important philosophical differences of opinion emerged during the Smog Check debate. By the early 1990s, California already had more than 30 years of experience pioneering air pollution controls. California's leadership rebelled at the notion that they did not have the capability to, on their own, design an effective I/M program tailored to the state's needs. Legislators viewed Smog Check as a states' rights issue, opposing what they called the one-size-fits-all mandates of the federal government. The Clean Air Act language lent credence to their case. The act instructed EPA to establish an enhanced I/M *performance standard* that states were to meet. California argued that a performance standard implied programmatic flexibility, as long as the ultimate emission reduction goal was reached. The final stages of the debate would circle back to this philosophical difference over the specificity of federal mandates: should the federal government mandate specific actions or allow states flexibility? In Chapter 10, where the E-REGS tool is introduced, the issue of prescriptive mandates is explored in greater depth by contrasting situations favoring command-and-control regulations (prescriptive means-oriented mandates) with those favoring performance-based requirements (flexible ends-oriented mandates).

### Sanctions

Congress gave EPA a mandate to punish states that failed to comply with the 1990 CAAA. The most potent punishment was the withholding of federal highway funds in response to a Clean Air Act implementation failure. During 1993, when the California economy was in recession, EPA threatened the state with the loss of nearly $1 billion in federal funds for opposing federal regulations.[6] State legislators did not believe EPA would follow through on its sanctions threat, given the state's political importance, air quality control history, and economic situation. As the debate progressed, EPA's threatened use of sanctions would further cement California's opposition to the agency's I/M program.

### Bad Timing: The Transition between Presidents George H.W. Bush and Bill Clinton

The timing of the Smog Check debate was very awkward. The seeds for the conflict were sown in the opening days of the administration of newly elected president Bill Clinton, when EPA career staff made critical I/M policy decisions as the new administration was coming up to speed on important issues. Compounding the timing problem was the fact that it would be many months before the Clinton administration appointed EPA leaders in Washington, D.C., and San Francisco. Thus, controversial EPA decisions were made prior to the arrival of key political appointees. Once the controversy blossomed into a full-scale state versus federal government battle, the intractable situation was handed to appointees as they first arrived to serve at EPA.

## A NOTE ABOUT THE DEBATE PARTICIPANTS

Readers may find themselves sympathizing with various viewpoints, from state, federal, or other perspectives, and perhaps criticizing actions taken by some debate participants. A few words of caution are necessary to temper reactions that might cause a reader to unfairly judge those involved in the story. The senior EPA career officials involved in the story were agency leaders who, over many years, had demonstrated their ability to simultaneously manage numerous people and complex issues within the confines of the federal bureaucracy. They also provided professional balance in a work environment where their politically appointed leaders came and went as federal administrations changed. California's participants included seasoned politicians and agency staff with years of regulatory experience. The politically appointed leaders at EPA had track records of substantial accomplishments before and after Smog Check.

Two examples serve to illustrate the skills and experiences of the individuals involved. As the debate unfolded, Mary Nichols became President Clinton's appointed EPA assistant administrator for air quality issues. Before Smog Check, Nichols served as the chair of CARB. Following Smog Check, Nichols and EPA administrator Carol Browner oversaw one of EPA's most far-reaching air quality initiatives: the 1997 establishment of new National Ambient Air Quality Standards (NAAQS) for ozone and fine particulate matter ($PM_{2.5}$). EPA implemented the new NAAQS following legal challenges that went to the U.S. Supreme Court. The 1997 ozone and $PM_{2.5}$ NAAQS decision set a new benchmark for U.S. efforts to curb air pollution and were a tremendous accomplishment for the Clinton–Browner EPA. At the time of this story, however, those achievements were still in the future. Following her service in the Clinton-era EPA, Nichols went on to serve as the cabinet secretary for the California Environmental Protection Agency, director of the University of California–Los Angeles (UCLA) Institute of the Environment, and later, once again, the chair of CARB.

During the debate, Dick Wilson headed EPA's Office of Mobile Sources (OMS), which bore responsibility for implementing enhanced I/M. Wilson had headed OMS for a decade prior to Smog Check; after Smog Check, he became deputy to Assistant Administrator Mary Nichols, and when Nichols left the EPA, Browner chose Wilson to replace Nichols as the acting assistant administrator. Following passage of the 1990 CAAA, Dick Wilson and his staff in Washington, D.C., and Ann Arbor, Michigan, found themselves at the center of a vortex: it was their job to implement the new act's most challenging mandates. For example, besides I/M, OMS weathered fuel reformulation efforts that sparked public outrage over perceived odor, health, and vehicle performance impacts. OMS implemented the Clean Air Act's trip reduction mandate to force businesses to encourage ridesharing, transit use, and other commute-to-work choices that would minimize drive-alone car use—a mandate so politically unpalatable that President Clinton signed its repeal into law on December 23, 1995. Public support was limited at best for programs like I/M and the other 1990 CAAA provisions that targeted motor vehicle use, as illustrated by press reports such as this: "As long as the fight for cleaner skies focused on big smokestacks, few people cared. . . . But clean-air rules are now biting into everyday life because they tackle a broader range of pollution problems . . . the most intense backlash is over intrusions on people's driving habits" (*Wall Street Journal* 1995).

Thus Dick Wilson and his team were charged with implementing the act's least popular requirements—and they were tasked with doing so on legal schedules many acknowledged as unrealistic (e.g., Portney 1990; Kraft and Vig 2003). For Mary Nichols, Dick Wilson, and the other debate participants profiled here, Smog Check represents only one policymaking experience, albeit an odyssey of huge and difficult proportion, within careers that accrued many important policy accomplishments.

Readers should also realize that the federal and state debate participants generally operated with the good-faith intention of reaching the same objective: improving air quality by using the least costly, most effective tools. Although economic and political interests loomed large throughout the debate, there was, at base, fundamental disagreement between the state and federal governments over how to inspect cars, as well as over how to set and implement policy. As the story unfolds, these core differences fueled the controversy and illuminate the case study presented here.

## BOOK ORGANIZATION

The rest of this study presents historical background leading up to the dispute (Chapter 2), an account of the debate (Chapters 3 through 5), a brief visualization of the debate period using timeline graphics (Chapter 6),[7] a look at the legacy of the Smog Check debate in terms of real-world outcomes in California and elsewhere (Chapter 7), an investigation into why the conflict occurred (Chapter 8), a discussion of the roles of the public and the press in environmental policy (Chapter 9), an examination of lessons learned and the introduction of E-REGS (Chapter 10), and an epilogue with valuable insights from several former career and appointed EPA officials (Chapter 11). Appendixes offer supplementary information about statutory language, motor vehicle emissions, and state experiences with RSD.

## NOTES

[1] Appendix A includes these requirements as originally specified in section 182(c)(3) of the 1990 CAAA.

[2] In December 2007, under the George W. Bush administration, EPA denied California's request for authorization to implement the state's new GHG emission standards, launching another California-EPA conflict. That conflict was resolved during the Barack Obama administration.

[3] Historically, a large fraction, typically more than 90 percent, of the total HC emitted by motor vehicles has been a more reactive class of compounds known as volatile organic compounds (VOCs). VOCs are sometimes mentioned interchangeably with HCs in the text.

[4] Appendix B includes U.S. and California emission standards.

[5] Various cost data emerged during the debate. At one point, for example, EPA estimated that IM240 equipment would cost only $106,000 and ASM equipment would cost $41,000 (Pidgeon et al. 1993).

[6] Dollar amounts cited in this study are for the year in which they are discussed. For example, the statement "EPA threatened the state with the loss of nearly $1 billion" implies 1993 US$.

[7] While reading the debate narrative in Chapters 3–5, some may find it helpful to refer to Figure 6-1 in Chapter 6, which presents a graphic summary of key events that occurred during the months leading up to and including the debate period.

CHAPTER 2

# A BRIEF HISTORY LEADING UP TO THE CONFLICT

To start the Smog Check story in early 1993, as EPA decisions began to track irrevocably toward a conflict, would be a bit like joining a conversation already in progress. Therefore, this chapter provides historical context before the next chapter delves into the controversy. It begins with a brief history of U.S. air pollution control efforts prior to the debate,[1] conveying the importance of car emissions and detailing some of the difficulties encountered in reducing them. As seen from the vantage point of history, the automobile was for many decades the central contributor to urban smog. The chapter then looks more closely at the history of the U.S. inspection and maintenance (I/M) experience before the conflict and examines the diverging policy viewpoints that marked the start of the debate. The underlying premise behind I/M was relatively simple: find polluting cars and require their repair. History shows, however, that meeting I/M's goal proved elusive.

## U.S. AIR POLLUTION CONTROL

### Late 1800s–1940

The origin of most U.S. air pollution problems from the late 1800s through the early twentieth century can be summarized in one word: coal. With the widespread use of coal, smoke and ash came to characterize urban America. As one U.S. historian recounted, "After the Civil War, individual cities often bragged about their prosperity and progress by advertising how many factory smokestacks and home chimneys threw plumes of coal smoke into their air. ... Home-heating furnaces, also coal-powered, had come into use. ... Coal—and its byproducts—literally permeated life in the United States" (Fifer 2002, *2*). In the early years of the twentieth century, coal's dominance as a U.S. fuel source meant that "thick clouds of black smoke from the combustion of high volatility bituminous coal remained a persistent feature of the industrial urban landscape" (Bachman 2007, *655*).

Similar smoke problems occurred in other countries as well, and one of the key air pollution events in the early twentieth century took place not in the United States, but in a European river valley. In early December 1930, a dense fog covered much of Belgium, including the valley along the river Meuse between the towns of Liège and Huy. The area was heavily industrialized, with metal and glass works and fertilizer and explosives plants. A strong temperature inversion had capped dense, cold, foggy air inside the Meuse Valley and prevented pollution from escaping. Of the valley's 9,000 residents, some 60 died within three days. In a 1931 report to Belgium's Royal Academy of Medicine, those who investigated the catastrophe established, for the first time, a link among temperature inversions, fog, and air pollution and noted that industrial and residential coal burning was to blame. Seventy years after the disaster, Belgium researchers stated that the Meuse Valley episode "provided incontrovertible evidence that air pollution could kill" (Nemery et al. 2001, *706*).

In the United States, Raymond Tucker, a professor at Missouri's Washington University, offered pivotal advice on overcoming smoke problems related to coal and led efforts from the mid-1930s to 1940 to create a St. Louis regulation that governed coal use and dramatically improved air quality. Although pollution problems and control efforts varied as a result of the Great Depression and World War II, Tucker's success in St. Louis led over time to other urban efforts and substantial progress toward reducing U.S. coal smoke, especially as natural gas and oil overtook coal in importance (Bachman 2007).

## 1940–1950

As U.S. smoke problems began to abate, a new problem emerged. In many areas of the United States, including Southern California, the response to World War II spurred increased growth in industry, the number of automobiles, and the emergence of smog. A report on pollution prepared for the California State Assembly described some of the health effects:

> In the Los Angeles region the increasing pollution of the atmosphere of the entire county reached a critical point in 1943. At that time, citizens began to experience the unusual phenomenon of eye irritation. ... It is first a feeling of dryness of the eyeballs often followed (possibly after rubbing of the eyes) by copious tears and inflammation. Many persons report a general feeling of nausea during intense smog periods and it is not uncommon for business offices to suspend operations and close-up shop during those times." (California State Assembly 1951, *22–23*)

In 1946, the *Los Angeles Times* hired St. Louis's Tucker to help diagnose and address Southern California's smog. Tucker determined that the smog, in contrast to coal-driven smoke problems, involved multiple sources, and warned that there was no single solution. The following year, California passed one of the first important U.S. air pollution laws to move beyond smoke problems: the state authorized counties to create air pollution control districts and regulate smog at the local level. Soon after, an air pollution catastrophe struck the town of Donora, Pennsylvania, located in the Monongahela River valley southeast of Pittsburgh. In late October 1948, during an episode with unfortunate similarities to the Meuse Valley disaster, 20 people died and nearly half of Donora's 14,000 residents were made ill or hospitalized when an unusually strong temperature inversion trapped cold air inside the valley, along with pollution from the town's steel and zinc plants. The ongoing problems in Southern California, plus the Donora catastrophe, heightened U.S. concern over air pollution. By 1949, researchers concluded that industrial processes, automobile and truck exhausts, and combustion were to blame for Los Angeles's smog problem, the country's most severe (SRI 1949; SCAQMD 1997).

Surprisingly, a November 1949 football game in Northern California—Washington State University versus the University of California–Berkeley—furthered the scientific understanding of smog. Thousands of fans at the game experienced the same intense eye irritation that had beset Los Angeles residents since the early 1940s. Although the entire San Francisco Bay Area had experienced a week of smoggy conditions prior to the game, the eye irritation occurred only in Berkeley and only on game day. Later research could find no industrial sources to blame; the unique factor was the thousands of automobiles caught in the Berkeley game-day traffic jams. A California State Assembly study concluded that automobile exhaust was the leading contributor to Berkeley's severe eye irritation episode. It was then short work to draw parallels to Los Angeles County, which had the "greatest concentration of motor vehicles in the world" (California State Assembly 1951, *41*).

## 1950–1970

If there were an air quality hall of fame, one of its first inductees would be A.J. Haagen-Smit, a California Institute of Technology scientist who launched the modern smog control era. In 1950, Haagen-Smit identified ozone as the key component of Los Angeles smog, and by 1952, he described that its atmospheric formation involved hydrocarbons (HC) and oxides of nitrogen ($NO_x$) emitted by oil refineries and automobiles. No sooner had Haagen-Smit's groundbreaking findings become available than another air pollution disaster struck: in December 1952, some 4,000 Londoners died when a temperature inversion trapped cold air and coal pollution close to the ground. Severe smog episodes soon followed in Southern California, and in light of the Donora and London catastrophes, U.S. concerns over air pollution continued to mount.[2]

Beginning in 1955, the U.S. federal government began to sponsor air pollution research and training. It would still be several years, however, before regulators acted to reduce car emissions. As an early air pollution expert lamented in a speech titled "The Dismal Future of Smog Control":

> During the war years the local public health officers of the cities and [Los Angeles] county fought and studied the thing [smog] under the most difficult conditions until at the end of 1946 they had concluded that the major source of smog was the automobile. ... [I]t is true that at Los Angeles the petroleum industry has been put to great expense to control every possible source of evaporation or leakage of hydrocarbons ... but even so, the simple fact of increasing population and increasing automobile use will more than counterbalance this removal. (Geraghty 1955)

Saddled with the nation's most severe smog problem, California eventually was the first in the United States to take steps to regulate automobile emissions. California required model year (MY) 1963 and later vehicles to reduce emissions from a car's crankcase and, beginning with MY 1966, mandated the nation's first-ever exhaust standards for HC and CO. Other parts of the United States also were experiencing smog problems during this period; for example, New York experienced bouts of heavy pollution in the early 1960s, and a particularly severe episode affected much of the Northeast in late 1966. Shortly afterward, Congress passed the 1967 Air Quality Act, establishing national air quality management procedures and preempting state actions to control automotive emissions. However, in recognition of California's pioneering work, the 1967 statute enabled the state to independently regulate motor vehicles—establishing a precedent that, for decades to come, would distinguish California's air pollution control efforts from those of all other states. Following in California's footsteps, the federal government required vehicles in the other 49 states to meet HC and CO exhaust standards beginning in 1968.

## 1970–1977

The year 1970 marked a turning point in the history of pollution control. The first Earth Day occurred in April, President Richard Nixon created the U.S. Environmental Protection Agency on December 2, and the Clean Air Act (CAA) was enacted by year-end. The new act provided the first comprehensive effort to set and attain air quality goals. It mandated creation of state implementation plans (SIPs) that were blueprints showing how each state would attain air quality standards. The legislation gave states five years to meet the standards, although EPA could grant a two-year time extension.

Beginning with MY 1971 vehicles, California instituted the first-ever $NO_x$ emission standards in addition to those already established for HC and CO. The federal government later instituted a similar requirement.

In February 1972, California submitted its first SIP to EPA. Unfortunately, the state found it impossible to demonstrate that Los Angeles could meet federal air quality standards by 1975. Under the terms of the 1970 CAA, EPA was forced to disapprove the Los Angeles area SIP, which it did in May 1972. Thus began a pattern repeated several times in the years that followed: because the SIP was inadequate to meet federal goals, public interest advocates sued EPA to, as required by the CAA, prepare a federal implementation plan (FIP) to correct the defective state strategy. In 1973, EPA prepared its federal response.

Governed by the act's unrealistic deadlines, EPA's Los Angeles FIP included extreme traffic control measures and anticipated gas rationing by the summer of 1977. Nor was the Los Angeles plan unique. As one researcher noted, "Pollution levels in Los Angeles, New York, Denver, Boston, and other major metropolitan areas were so high that these cities could not meet air quality standards without imposing such unpopular measures as parking bans and surcharges, mandatory bus and carpool lanes, and gasoline rationing of up to 90 percent" (Melnick 1983, *300*). Given that so many areas were unable to demonstrate attainment of the air quality standards by the CAA deadline, litigants compelled EPA staff in 1973 to write approximately 25 FIPs focused on transportation control. The plans were unpopular forays by the federal government into land use and transportation actions traditionally reserved for local government. One of EPA's Boston controls, for example, required all firms with 50 or more employees to reduce their parking spaces by 25 percent. The opposition to these efforts made clear how difficult it would be to control driving behavior as a method of reducing emissions (Melnick 1983; U.S. EPA 1990).

As EPA and the states struggled with the 1970 CAA deadlines, regulations forced automakers to reduce new-vehicle emissions. Catalytic converters were introduced in MY 1975, facilitating substantial emission reductions from new vehicles. To prevent catalyst poisoning, as well as to reduce exposure to airborne lead, lead-free gasoline became available in 1975 to fuel the new catalyst-equipped cars.

By 1976, it was apparent that EPA's transportation control FIPs generated fierce opposition and, if implemented, would result in unacceptable social and economic consequences. As a result, the agency changed its policies. In Los Angeles, for example, the release of EPA's FIP triggered lawsuits to block its implementation, and EPA officially withdrew the plan in 1976 (U.S. EPA 1990).

## 1977–1987

The decade from 1977 to 1987 marked the second major attempt by the United States to achieve clean air in urban areas. With the failure of the 1970 CAA, Congress enacted the 1977 Clean Air Act Amendments (CAAA) and gave states an additional five years—until December 31, 1982—to meet air quality standards for urban-scale pollutants such as CO and ozone (then referred to as photochemical oxidant). States that were unable to demonstrate attainment by 1982 could opt to take an additional five years, until December 31, 1987, provided they implemented I/M programs. The 1977 CAAA also broadened California's independence. As stated by EPA, "The 1977 amendments to the CAA expanded the flexibility granted to California in order 'to afford California the broadest possible discretion in selecting the best means to protect the health of its citizens and the public welfare'" (U.S. EPA 1990). In response to the 1977 CAAA, California updated its SIP. Despite the inclusion of more than 100 control measures, however, the SIP could not demonstrate how the Los Angeles area would attain the CO or ozone standards by 1982, or even 1987. Many other U.S. areas were also unable to meet the new deadlines. The air quality problems were simply too severe to solve within the time parameters set by Congress.

In 1984, as U.S. urban areas worked unsuccessfully to meet the 1977 CAAA deadlines, a major air pollution disaster struck India. Shortly after midnight on December 3, a Union Carbide fertilizer plant in Bhopal, India, leaked methyl isocyanate gas into the surrounding neighborhood, killing approximately 3,800 people and injuring well over one hundred thousand others. The disaster, originating from a U.S.-owned corporation, increased international concern over air pollution and the handling of toxic substances.

In October 1987, with the CAAA's December 31 deadline for clean air fast approaching, U.S. EPA administrator Lee Thomas noted that the ground-level ozone problem had proven "particularly intractable." In an article titled "Next Steps in the Battle against Smog," Thomas reflected on the emission reduction progress the country had made to date and offered advice for the next statutory attempt at clean air. He suggested setting new clean air deadlines that were "realistic and tailored to the circumstances of individual nonattainment areas." And "most important of all," observed Thomas, "the national strategy should allow the states substantial latitude in designing attainment plans that are efficient, effective, and politically acceptable. We at EPA are committed to the attainment of the ozone standard nationwide. But we recognize that, in some areas, achieving that goal could require extraordinary control actions that may be costly, socially disruptive, and politically unpopular" (Thomas 1987, *4*).

## 1987–1990

Once 1987 drew to a close, U.S. clean air efforts fell into unclear statutory territory. The CAA's deadlines had come and gone, and Congress had yet to pass new legislation. In the absence of a new law, litigation helped maintain momentum to reduce pollution. In January 1988, acting in response to lawsuits, EPA disapproved California's SIP as being unable to demonstrate attainment of the CO and ozone standards as required by the end of 1987. Then, in February 1988, California environmental advocates sued EPA to compel creation of a FIP to achieve clean air. By late 1988, the U.S. air pollution picture looked far worse than it had in many years. During that year, the United States experienced record heat and increased ozone levels, and dozens of cities were added to the list of areas failing to meet ozone standards (Waxman 1992).

In 1989, scientists published a landmark study that began to reshape understanding of motor vehicle pollution. The scientists used a tunnel that ran under the Van Nuys Airport in Southern California as a real-world test chamber in which to measure automotive pollution. Researchers linked the cars driving through the tunnel to measured pollutant concentrations, and then calculated vehicle emissions. When the researchers compared their tunnel results with predictions generated using then-state-of-the-art EPA and California computer models, they found a startling result: the computer models, essential tools needed to prepare SIPs, underpredicted automotive emissions by factors of two to four.

The findings were so important that they triggered several follow-up studies, which also found that the models were alarmingly wrong (Ingalls et al. 1989; Pierson et al. 1990).

Despite the absence of an updated federal law, the Los Angeles area's air pollution agency, the South Coast Air Quality Management District (SCAQMD), continued forward with a strategy to reduce Southern California pollution. In March 1989, SCAQMD released an air quality management plan that had been under development for five years. The new plan sought to meet CO standards by 1997 and ozone standards by 2007. Because of the unique severity of the Los Angeles pollution problem, the plan acknowledged that extraordinary reductions in motor vehicle emissions were needed, and that meeting long-term ozone goals required the development of new technologies (U.S. EPA 1990).

In September 1990, to promote the creation of new-technology vehicles, the California Air Resources Board (CARB) established a visionary Low Emission Vehicle (LEV) Program, which mandated reducing more than 99 percent of the HC exhaust from new cars.[3] The program's ambitious emission targets helped pave the way for automakers to introduce in the United States, nearly a decade later, the first commercially available hybrid-electric vehicles.

## 1990–1992

Following several years of negotiation, Congress substantially restructured the CAA, and on November 15, 1990, President George H.W. Bush signed the 1990 CAAA into law.[4] The text of the Clean Air Act grew sixfold from 1970 to 1990, moving from broad goals to detailed mandates and deadlines for EPA, industry, and states. The new law scaled clean air deadlines to match the severity of a region's problem; mandated automotive emission standards comparable to those already adopted by California; acknowledged the need for advanced, though as-yet-unknown, technology to solve Southern California's ozone problem; and established the need for enhanced I/M programs in the nation's worst-polluted regions.

As federal and state regulators digested the 1990 CAAA's new requirements, scientists continued to examine the mismatch between real-world pollution measurements and car emissions predicted by computer models. By 1992, the National Research Council (NRC) acknowledged that recent scientific findings "conclusively demonstrated" that EPA's and California's computer models underpredicted automotive emissions (NRC 1992, *300*). In June 1992, EPA published a report that acknowledged and diagnosed the computer problems, and agency scientists worked with others to run further experiments in Baltimore's Fort McHenry Tunnel (U.S. EPA 1992a, 1992c).

One of the many challenges involved with regulating cars was that as they aged, their emission controls became less effective because of worn-out parts, poor maintenance, or outright tampering by owners. By 1992, EPA estimated that on average, in-use cars emitted three to four times the pollution of new cars (U.S. EPA 1992e).

In 1992, the NRC released a landmark publication, *Rethinking the Ozone Problem in Urban and Regional Air Pollution.* In it, the NRC pointed out that since 1970, the United States had tried and failed three times to meet ozone health standards: the 1970 CAA had set 1975 as the first clean air deadline; when that deadline was missed, the 1977 CAAA gave states until 1982 to reduce ozone; then areas failing to meet that milestone were given until 1987—a third deadline missed by about 100 areas. The NRC found that these failures were due in part to the fact that scientists and regulators had underpredicted automotive pollution and overpredicted the benefits of I/M programs implemented to date. In addition, ozone control programs had relied too heavily on reducing HC emissions alone; future efforts would have to reassess the benefits of more aggressive $NO_x$ controls. The report also noted that based on 1985 national emission estimates, transportation sources were the single largest contributors to the nation's HC and $NO_x$ emissions (NRC 1992).

As 1992 drew to a close, the United States was embarked on a renewed effort to control air pollution. Much as they had been at the outset of the smog problem 50 years earlier, cars remained the key problem, as well as the focus of regulatory attention.

## A BRIEF HISTORY OF I/M

### 1960–1990

Similar to its pioneering role with new-vehicle emission standards, California also helped initiate the use of vehicle inspections to identify problem polluters. Beginning in the mid-1960s, California instituted a statewide change-of-ownership motor vehicle inspection requirement. Private garages certified by the state—known as Blue Shield stations for their distinctive logo—inspected vehicles before the owners could sell them. Regulators reasoned that inspections would identify polluting vehicles, and then the owners would be required to repair problems (CA IMRC 1993).

Later, the 1970 CAA included I/M as a control option, and, in 1974, New Jersey created the first routine U.S. I/M program; other states followed in the mid-1970s. The New Jersey program was a centralized I/M system, with an annual test requirement rather than a change-of-ownership requirement as in California. New Jersey used a simple tailpipe test that measured pollutants while a vehicle idled its engine; this test, and others like it, became known as idle tests (U.S. EPA 1992d).

The 1977 CAAA required states to submit air quality plans to EPA that provided, "to the extent necessary and practicable, for periodic inspection and testing of motor vehicles to enforce compliance with applicable emission standards." In addition, areas unable to achieve ozone or CO standards by 1982 had to "establish a specific schedule for implementation of a vehicle emission control inspection and maintenance program." In response, various states implemented I/M programs.

Beginning in 1979 and continuing through March 1984, California operated a limited centralized I/M program in the Los Angeles area. Seventeen stations operated by a private contractor, Hamilton Test Systems, tested change-of-ownership vehicles. Failing vehicles were sent for repairs, and motorists could choose to repair vehicles at any private garage licensed by the state Bureau of Automotive Repair (BAR). The centralized program was found to be much more effective than the Blue Shield program, which suffered from poor inspection quality and inadequate repair work. Public acceptance of the centralized program was limited, however, with many participants complaining about the inconvenience of having to go to a test center and then get their vehicle repaired elsewhere. The program was also criticized for failing to provide an adequate number of testing facilities, given the size of the Los Angeles region (U.S. EPA 1992c; CA IMRC 1993).

Given the unpopularity of I/M, the California legislature refused at first to augment its Los Angeles I/M program to include other regions and meet the federal mandate in the 1977 CAAA. In response, in December 1980, EPA sanctioned California by delaying federal highway and sewer project funding (CARB 2000b).

The use of sanctions helped motivate California to expand its I/M program, and in 1981, California state senator Robert Presley introduced Senate Bill (SB) 33. The legislation was initially written to expand the Los Angeles area's centralized program to include the state's other regions. Political opposition to centralized testing, however, led Senator Presley to amend the plan to require garage-based inspections, an approach ultimately agreed to by EPA (CA IMRC 1993). This early, and apparently effective, use of sanctions established a precedent for EPA career staff, many of whom later became involved in the California-EPA dispute.

Passage of SB 33 in 1982 created California's first routine inspection program. Named Smog Check and scheduled to begin in March 1984, the program applied to polluted regions of the state and required all gasoline-powered cars and light trucks registered in those regions to pass an inspection once every two years and on change of ownership (CA IMRC 1993).

Despite initial forecasts of success, Smog Check performance failed to meet expectations. SB 33 included requirements to analyze how well the program performed, and CARB and BAR joined forces over several years to randomly recruit vehicles from the program and monitor their emissions. By April 1987, when the state's official I/M Review Committee (IMRC; then composed of air quality management professionals but later restructured by the legislature) delivered its first Smog Check evaluation report, it was apparent the program was achieving only half the hoped-for HC and CO emission reductions. Studies showed, for example, that Smog Check stations missed engine-tampering problems, and in the Los Angeles area, some vehicle owners performed pre-test maintenance followed by post-test engine adjustments that reduced program impacts. One study found that of vehicles that had received a passing Smog Check test, more than half failed a roadside test six months later (e.g., Lawson et al. 1990; Pierson 1996). In response, the state legislature mandated program improvements to take effect January 1, 1990.

By December 1991, the California IMRC had evaluated the improved Smog Check program and continued to find disturbing results. Vehicles randomly stopped and tested at the roadside by state program auditors failed the Smog Check test at a rate 50 percent greater than the failure rate at Smog Check stations. The IMRC, reporting to the state legislature, suggested that centralized testing might ultimately be needed to achieve emission reduction goals (CA IMRC 1993).

## I/M and the 1990 CAAA

The 1990 CAAA consisted of 300 pages of fine-printed, double-columned text. The enhanced I/M program requirement, which, before the end of the California-EPA I/M conflict, would threaten to undo the entire statutory framework, took up a single page amid the law's myriad mandates. However, it was EPA's view, which the agency shared in numerous presentations and in its 1992 I/M regulations, that enhanced I/M was not only one of the most effective air pollution control programs known, but also the single most cost-effective measure within the entire 1990 CAAA. EPA noted that enhanced I/M was 7 times more cost-effective than new tailpipe exhaust standards for automobiles, and at least 10 times more than additional controls on small and large stationary sources of air pollution (U.S. EPA 1992e). The agency was not alone in its assessment. The California IMRC reported that "enhancement of the Smog Check program is the single most effective air pollution control action currently available" (CA IMRC 1993, *27*).

Congress established enhanced I/M as a requirement for the worst-polluted CO and ozone regions and set two goals to be met by November 15, 1992. First, states had to give EPA a plan describing how they would create their enhanced programs. Second, they needed to implement those programs and "comply in all respects with guidance published in the Federal Register (and from time to time revised) by the [EPA] Administrator for enhanced vehicle inspection and maintenance programs" (Appendix A includes the 1990 CAAA enhanced I/M requirement).

Note that Congress allowed no time between the deadline to submit a plan and the deadline to implement a program. (EPA later attempted to rectify this problem by modifying the implementation deadlines included in its regulations.) In addition, lawmakers required EPA to give states guidance on how to create and implement programs. The guidance was effectively a requirement for EPA to issue regulations, as states needed to comply with the guidance "in all respects." Congress said EPA's guidance should include the following:

- a *performance standard* achievable by a program combining emission testing, including on-road emission testing, with inspection to detect tampering with emission control devices . . . ; [and]
- program administration features . . . to attain and maintain the performance standard. (emphasis added)

Lawmakers also defined required program elements, including annual, centralized emission testing, unless states demonstrated that biennial or decentralized inspections were as effective. Before states could design and implement programs, however, EPA needed to release its guidance. And that, unfortunately, was the seed from which the California-EPA Smog Check debate sprouted.

## The EPA I/M Regulatory Package

The 1990 CAAA effectively required EPA to issue I/M regulations and help states implement enhanced I/M by November 15, 1992, just two years after the new amendments became law. Anticipating that EPA was falling behind schedule, the city and state of New York, along with an environmental organization called the Natural Resources Defense Council (NRDC), sued EPA to issue I/M regulations. In response, EPA received a court-ordered mandate to publish final I/M regulations by November 6, 1992 (Hurley 1992).

To prepare its I/M rule, EPA published a Notice of Proposed Rulemaking on July 13, 1992, and solicited public comment. Given the conflict that erupted later, California's comments on the proposed I/M rule are especially interesting. On August 26, appointees of California governor Pete Wilson, including

California EPA secretary James Strock, wrote, "We urge EPA to delete the one-more-try-at-test-and-repair program option from the final rule." However, foreshadowing the debate to come, the state officials also stated: "New legislation will be needed to authorize the enhancements. Our success in obtaining this legislation, and implementing the needed improvements to our program, depends on our ability to design an enhanced program which addresses concerns of our legislature. . . . As a result, California's enhanced I/M program may differ considerably from the program design EPA has so strongly advocated." The state closed its comments with a request that, among other things, EPA give "highest consideration" to reducing the overly prescriptive content of its regulations (Strock and Poat 1992).

Spurred on by state comments supporting test-only facilities, and ignoring calls for less prescriptive rules, EPA embraced the opportunity to break from past problems and took full advantage of statutory language requiring "operation on a centralized basis, unless the state demonstrates to the satisfaction of the EPA Administrator that a decentralized program is equally effective." When EPA published its final I/M regulations on November 5, 1992, one day before the court-ordered deadline, it created a regulatory scheme that virtually abolished programs where a single garage tested, repaired, and certified vehicles. The regulations also created a firestorm of problems for EPA and the states.

EPA's court-ordered enhanced I/M regulations marked a clear departure from past I/M policies. The regulations included numerous provisions and required the use of new technologies, such as dynamometers, a 240-second test procedure that simulated real-world driving (called the IM240), and evaporative emission tests.[5] The regulatory centerpiece, however, was EPA's move to test-only programs. Drawing lessons from programs established since the mid-1970s, EPA's automotive specialists in Ann Arbor, Michigan, determined that decentralized, garage-based inspection programs, such as California's, were vastly inferior to centralized, test-only systems already operating in numerous states.

Years of field experience contributed to the development of EPA's position. From the late 1970s until the early 1990s, EPA conducted motor vehicle tampering surveys, which involved pulling over randomly selected vehicles to the roadside to complete emission measurements in the field or selecting vehicles from inspection test lanes for further assessment. The surveys identified whether vehicles had malfunctioning emission control equipment, due either to equipment failure or deliberate tampering, and covered tens of thousands of vehicles around the United States. EPA used the findings to assess whether I/M programs operated properly (NRC 2001). In addition, from 1984 through 1992, EPA completed about 100 I/M program audits. The surveys and audits generated some negative observations. EPA noted that vehicles participating in decentralized (garage-based) programs exhibited emission control problems at rates 20 to 50 percent higher than vehicles from centralized programs, and covert audits showed that vehicles in decentralized programs improperly passed inspections 30 percent of the time in California and 34 to 82 percent of the time elsewhere. Although EPA acknowledged that centralized systems could also experience these problems, the agency maintained that it was "virtually impossible to improperly test a vehicle for tailpipe emissions" in a centralized setting (U.S. EPA 1992e). Later investigations would cast doubt on EPA's technical findings; in 1992, however, the agency's assessment stood as one of the few available evaluations of the U.S. I/M experience.[6]

In brief, EPA reasoned that service stations had incentives either to falsely fail vehicles in order to sell unnecessary repairs or to falsely pass cars to maintain good business relationships with regular customers. Either way, service stations had a conflict of interest when performing inspections. In addition, EPA staff believed that service station technicians inspected cars infrequently and therefore lacked expertise. EPA concluded that centralized I/M programs were simply more cost-effective, and that better inspections were performed when the "testing agent did not have any interest or involvement in the repair of vehicles." The agency noted in its final regulation that, in response to EPA's proposed rule, "state governments made it clear that they saw no way to achieve the [enhanced I/M] performance standard with a test-and-repair system" (U.S. EPA 1992e).

## Flexibility for States to Design Their Programs

Although the final I/M regulations led states down a narrow path toward implementing a test-only program, the regulatory package included language that, at first glance, gave states ample flexibility to design the program of their choice. EPA referred to a program that exactly matched its requirements as a "model program" meeting the agency's "performance standard": "States have flexibility to design their own programs *if they can show that their program is as effective as the 'model' program used in the performance standard*" (emphasis added). EPA defined its performance standard as the "minimum amount of emission reductions, based on a model or benchmark program design, which a program must achieve." The agency noted that "the concept of a performance standard provides state flexibility, as long as the numerical goal for emission reductions is attained. *A State may choose to vary any of the design elements* (except those required by the Act) of the model program *provided the overall effectiveness is at least as great as the performance standard*" (U.S. EPA 1992e, *52951a–52953a*; emphasis added).

States had to address two key questions in their quest to design and receive EPA approval for an alternative I/M program. First, what was the EPA performance standard? Second, how did an alternative program compare with the performance standard? EPA defined the performance standard in clear terms: measured against an area without an I/M program, enhanced I/M in a typical area had to achieve a reduction of 28 percent in volatile organic compound (VOC) emissions, 31 percent in CO emissions, and 9 percent in $NO_x$ emissions by the year 2000. These reductions were substantial, and achieving them would be the cornerstone of most air quality plans that included enhanced I/M. For example, California's ozone SIP ultimately relied on enhanced I/M to achieve one-quarter of all the new emission reductions sought by 1999 (CARB 2000b).

## The Insurmountable Hurdle: The 50 Percent Discount for Test-and-Repair Programs

Despite the appearance of an opportunity to demonstrate that an alternative program design met EPA's performance standard, states such as California faced an insurmountable hurdle to implement or improve a garage-based test-and-repair program. The catch was that a state had to demonstrate its proposed program's effectiveness using an EPA computer model that discounted by 50 percent the emission benefits of test-and-repair programs.

EPA had developed a computer model over many years that enabled analysts to predict motor vehicle emissions. The model, known as MOBILE, included hundreds of assumptions based on various data analyses and studies conducted by EPA and others. It also happened to be the same computer model that the NRC found had unreliably estimated vehicle emissions. In November 1992, EPA acknowledged that

> EPA's emission factor model (MOBILE) for I/M programs contains a set of default assumptions reflecting the fact that decentralized test-and-repair programs have in the past been significantly less effective than centralized programs with similar design features in finding and fixing emission problems. EPA believes it could not accept any of the currently operating decentralized programs as equally effective to centralized. With these effectiveness losses, *it is not possible for a decentralized test-and-repair program to meet the performance standard for enhanced I/M*, regardless of the test type or vehicle class coverage." (U.S. EPA 1992a, *52959a–b*; emphasis added)

The potential impact of the 50 percent discount can be illustrated using Sacramento. To meet ozone standards, Sacramento needed to reduce its 1990 HC emissions by approximately 40 percent. It committed to various controls, including enhanced I/M, to meet the HC target. Local air quality officials assumed that an enhanced Smog Check program would reduce HC emissions by 5 tons per day.

If EPA discounted those benefits by 50 percent, however, Sacramento would have to implement supplemental controls, at additional cost, to eliminate 2.5 tons of daily HC emissions.

In its final regulations, EPA stated that the 50 percent discount could be reduced if a state demonstrated to the agency's satisfaction that its test-and-repair system merited more emission reduction credit. However, in addressing public comments on its draft regulations, EPA responded that "neither EPA nor the states or other commenters know of any 'other program elements taken together'" that would allow test-and-repair programs to "achieve equal effectiveness" with a test-only system (U.S. EPA 1992a, *52974c*). The 50 percent discount, on its own, derailed state plans to implement garage-based programs.

## EPA Response to State and Industry Concerns: Jobs, Costs, Convenience

Anticipating opposition to its final regulations, EPA spent considerable time documenting that test-only programs would increase inspection and repair industry jobs, reduce motorist costs, and improve convenience for most of those seeking inspections.[7] Despite its best efforts to downplay and mitigate concerns over jobs, cost, and convenience, however, EPA weathered challenges on all three issues throughout the Smog Check debate. State legislators, who were sensitive to constituent concerns, contested the agency's data and assertions. The EPA staff did its best, in a short time, to prepare a scientifically defensible case for the agency's positions. Unfortunately, the abbreviated schedules in the 1990 CAAA, combined with a court-ordered deadline to publish regulations, exacerbated the agency's struggle to assemble and analyze data. As one longtime EPA career official observed, "Court deadlines are sacrosanct at EPA—we never miss them. The Ann Arbor staff lamented that they were unable to fully buttress the technical elements of the final I/M rule because they simply ran out of time." As the policymaking process unfolded, EPA's positions were countered by industry, academics, public interest groups, and air quality professionals with numerous alternative views.

## The Confluence: EPA Mandates vs California's Gold Shield Proposal

In November 1992, CARB officially committed to EPA that the state would implement an enhanced I/M program consistent with the Clean Air Act and EPA regulations. In its commitment, CARB informed EPA that it was "supporting the adoption of legislation in the 1993 California legislative session that would authorize enhancements to California's Inspection and Maintenance" program, and that the enhancements being sought would "meet or exceed" EPA's I/M program performance standards (Boyd 1992). State Senator Robert Presley, sponsor of the original 1984 Smog Check program, had agreed to sponsor the needed legislation. Presley, a Democrat, represented Riverside County, an area east and downwind of Los Angeles that often experienced the region's worst air pollution.

To meet EPA's required implementation milestones, California needed to pass authorizing legislation during the 1993 legislative session, design a new program during 1994, and begin phase-in of the new program by January 1, 1995. Decisions needed to be made quickly about the structure of a new Smog Check program, because Senator Presley planned to introduce a legislative package at the outset of the 1993 session, which began in January.

As of late 1992, the California IMRC was finalizing its upcoming fourth report to the legislature on Smog Check. The report included an assessment of the program's effectiveness and recommended a package of improvements to meet federal enhanced I/M requirements. The IMRC shared a draft of the report with EPA and asked the agency to formally comment on the state's proposed approach to improve Smog Check.

In its report, the IMRC found that the existing Smog Check program, the most expensive in the country, was about half as effective as an ideal program. Though the program had improved over the years, it still fell short of program goals and was substantially less effective than the EPA enhanced I/M performance requirements (CA IMRC 1993). The committee's findings were, if anything, overly generous about Smog Check's performance compared with the critiques of others. Some scientists expressed concern that Smog Check had little impact on real-world emissions, because some data showed emission reductions only for the day of the test.[8] Especially problematic from EPA's view was the policy direction in which the state appeared to be headed: the IMRC promoted a program with broader technical and political support than EPA's preferred test-only plan. The state's plan allowed for a continuation of at least some test-and-repair stations.

The IMRC believed that EPA fundamentally erred in requiring centralized testing programs paired with IM240 tests and argued that to ensure successful retests of failing vehicles, garages doing repair work needed to have the ability to duplicate the test being conducted at the testing center. Anything short of duplicating the test would involve guesswork about repair adequacy and leave motorists vulnerable to repeated failures at a centralized station. The IMRC believed that EPA requirements to use expensive IM240 equipment all but precluded a garage from being able to purchase equipment identical to that used by inspection centers. In addition, it disagreed with EPA's interpretation that decentralized programs merited only 50 percent of the emission reduction credit given to EPA's model enhanced I/M program. The committee acknowledged that the existing California program was only half as effective as what was required under enhanced I/M, but noted that program improvements such as dynamometer testing would surely increase program effectiveness (CA IMRC 1993).

To overcome Smog Check's problems, the IMRC recommended that instead of using EPA's IM240 procedure, the state adopt a far less expensive dynamometer-based test, the acceleration simulation mode (ASM) test. It also recommended creation of a hybrid program using so-called Gold Shield test stations. The concept involved first testing all vehicles at a test-only facility, then allowing failing vehicles to be repaired and retested at Gold Shield test-and-repair garage stations. Gold Shield stations would perform dynamometer tests and employ highly trained technicians. As a further safeguard, the committee noted that the worst-polluting vehicles, also called high emitters or gross polluters, and vehicles that had been tampered with would still be required to return to test-only facilities for final certification. The IMRC estimated that the hybrid program would cost $24 per test, a reduction over the existing program's $28 average cost (CA IMRC 1993).

In late 1992, EPA and California air quality professionals began earnest discussions to determine what program design EPA would find acceptable. As the next three chapters describe, what unfolded proved to be the undoing of the entire federal I/M program.

## NOTES

1 Several sources were especially helpful during the preparation of this discussion; they offer additional historical information for interested readers: Air Pollution Control Association (1982) chronicles a 75-year history of air quality management; Melnick (1983) assesses challenges EPA faced when implementing transportation controls following the 1977 CAAA; U.S. Environmental Protection Agency (1990) provides a history of southern California's SIP; South Coast Air Quality Management District (1997) gives a 50-year retrospective on fighting Los Angeles area smog; Davis (2002) describes the 1948 air pollution catastrophe in Donora, Pennsylvania; National Research Council (2004) assesses U.S. air quality management; Bachman (2007) outlines the evolution of U.S. health-based air quality standards.

2 A 1953 photograph illustrating the era's smog problems shows a woman wiping her irritated eyes while crossing a street in downtown Los Angeles. Across the street, only dimly visible through the smog, appears the outline of City Hall. For this and other historic smog pictures, see *Los Angeles Times* 2010.

3 The 1990 LEV Program required major automakers to have at least 2 percent of their MY 1998 sales fleet be zero emission vehicles (ZEVs), and to increase the ZEV fraction to 10 percent of the sales fleet by MY 2003. Later program modifications adjusted those targets and granted partial ZEV credits to vehicles such as hybrids. Appendix B includes federal and California automotive emission standards.

4 See Bryner 1995; Cohen 1995; and Waxman 1992 for discussions on the passage of the 1990 CAAA.

5 In the 1990s, regulators realized that perhaps half of hydrocarbon pollution from automobiles was from gasoline evaporation, with the other half from fuel combustion; as of 1992, evaporative emission testing had not yet been tried within I/M programs (U.S. EPA 1992e; Pierson et al. 1999).

[6] A later reassessment of EPA's 1985–1992 roadside data found little difference among vehicles from non-I/M, centralized I/M, and decentralized I/M areas (Lawson et al. 1995; NRC 2001).

[7] EPA's regulatory package included arguments largely supported by the California IMRC. *Jobs*: EPA said a more effective test better identified problems and generated more repair work. The IMRC estimated that California would gain 500 jobs by implementing test-only I/M. *Costs*: EPA estimated that biennial, centralized I/M would reduce emissions at a cost of up to $1,600 per ton of pollutant reduced. The IMRC estimated that a centralized IM240 program would reduce emissions for approximately $2,200 per ton. *Convenience*: Some policymakers were concerned that in a centralized system, failing vehicles would make at least two additional trips: for repairs and a retest. EPA argued that enhanced I/M reduced inconvenience by requiring tests every two years rather than annually, and estimated that 80 to 90 percent of vehicles would pass their first inspection and avoid multiple trips. The IMRC estimated that 70 to 80 percent would pass their first inspection and held that centralized I/M would, overall, be more convenient than garage-based I/M (U.S. EPA 1992e; CA IMRC 1993).

[8] Prominent among these findings were results later published in the peer-review literature (Lawson 1993, 1995); see also Pierson 1996 for a summary of findings.

# PART II
# THE SMOG CHECK CONFLICT AND ITS OUTCOMES

CHAPTER 3

# EPA POLICY RUNS INTO CALIFORNIA POLITICS

This chapter and the two that follow describe the I/M debate that took place between EPA and California in 1993 and 1994. The chapters are written from the perspective of those staff in EPA's California office responsible for negotiating with state officials. To provide as rich a case study as possible, the text reconstructs many of the key debate discussions involving EPA decisionmakers. These accounts are largely based on detailed notes I took at the time of the Smog Check debate, supplemented by EPA memoranda, newspaper accounts, and legislative debates. Only in rare cases were actual transcripts available of discussions held, such as when recordings were available from legislative debates. Where notes and recollections were able to support what I believe is a reasonably accurate reflection of the discussions, the text presents dialogue with the use of quotations. These quotes help convey the tenor of the discussions that took place throughout the policymaking process.

An extensive attempt has been made to portray discussions accurately as they occurred, and several debate participants reviewed and commented on the text to help ensure that it appropriately represents the essence of the policy dispute. Inevitably, however, there may be some departures from the exact language used by individuals in the quotations presented here. The point of the narrative is to reconstruct, for analytical purposes, the evolution of the debate, the real-world policymaking challenges faced by decisionmakers, and the problem-solving efforts that eventually led to a resolution. The detailed narrative also enables the extraction of meaningful lessons tied to real-world events; these lessons are discussed in later chapters.

## KEY DEBATE PARTICIPANTS

Many players' involvement in the I/M debate waxed and waned, but not so with EPA, which remained a key participant throughout. The U.S. Environmental Protection Agency is a regulatory machine composed of many far-flung parts. EPA policymakers were primarily in Washington, D.C.; my team was in San Francisco. We were responsible for working with states to implement and enforce federal law. The agency had 10 such regional offices; ours, Region 9, covered Arizona, California, Hawaii, and Nevada. In San Francisco, Dave Howekamp headed the EPA air quality program. Dave had spent virtually his entire career in the federal government, rising to oversee approximately 120 people responsible for implementing air pollution policy in the worst-polluted regions of the United States. Dave was at the top of the career ladder and reported to a political appointee, the EPA regional administrator. Working as part of Dave Howekamp's regional air program, I had recently left the private sector, and as the Smog Check debate began to unfold, I had just completed my first year as the Region 9 mobile sources section chief.

In addition to its headquarters and regional offices, EPA had laboratories and research centers scattered around the country. One of these, the EPA Office of Mobile Sources (OMS) (later renamed the Office of Transportation and Air Quality), was managed from Washington, D.C., but most of its staff was stationed in Ann Arbor, Michigan, near the Detroit-based automotive industry. OMS regulated pollution from motor vehicles, and the Ann Arbor team was responsible for developing and implementing EPA's I/M policies. A career official, Richard (Dick) Wilson, led OMS; he reported to the EPA assistant administrator for air and radiation, a political appointee responsible for setting air pollution policy throughout the United States. Of the many people on Dick's staff in Ann Arbor, several were especially critical to the I/M debate. Charles Gray was one of the lead EPA managers in Ann Arbor working on I/M; under Charles was Phil Lorang, responsible for I/M and other issues. One of the managers under Phil's program area was Gene Tierney, a section chief whose main responsibility was I/M.

Together, Dick, his staff of career scientists and engineers, and the regional mobile sources teams were responsible for implementing the Clean Air Act's I/M mandates.

In addition to career staff, several EPA political appointees were integral to the Smog Check story. At the outset of the debate, however, almost all of these leaders had yet to be identified. George H.W. Bush lost the November 1992 U.S. presidential election to Bill Clinton. During January 1993, Clinton appointees began to take the place of their Bush administration counterparts. The transition time was a critical period. Caught between a lame duck administration that issued the new I/M regulations and an incoming administration that had yet to name many of its environmental leaders, the career staff at EPA entered a rare policymaking window when key issues were addressed without the usual involvement by appointed leaders.

At the state level, numerous California participants were involved, including elected officials, career bureaucrats, and others. These state players will be introduced as the story unfolds.

To help readers differentiate among the key players involved in the Smog Check debate, the text generally refers to EPA officials by their first names and to virtually all other players by surname. The narrative unfolds from the perspective of the EPA regional office, with the interactions characterized entirely from this perspective, thus the use of first names for EPA staff. The exception is EPA administrator Carol Browner, referred to by her official title or last name.

## THE GREAT TRAIN WRECK: DECEMBER 1992–FEBRUARY 1993

By late 1992, the stage was set to improve Smog Check. The biggest unknown was whether EPA would approve the California Gold Shield scheme, a plan to have highly qualified service stations test and repair cars.

We in the San Francisco EPA office conferred with staff in Ann Arbor and Washington, D.C., to assess whether EPA would approve Gold Shield. Staff at those offices were cautious about where California was headed. As of late 1992, the Ann Arbor staff was willing to consider the plan, but only if Gold Shield stations tested a small fraction of cars. They worried that if EPA allowed California to run a sizable test-and-repair program, other states would abandon their efforts to create test-only systems.

During several brainstorming sessions, San Francisco and Ann Arbor EPA staff held telephone meetings to get policy direction from senior management in Washington, D.C. Our telephone call on December 8, 1992, was typical. Dick Wilson, head of OMS, asked Region 9 and Ann Arbor for input on whether EPA should approve California's proposals. Dick helped us think through what California was proposing: if all cars started at test-only, he asked, but some of the failing cars could be retested and certified at Gold Shield stations as long as they weren't high emitters, was this something we wanted? One of the Ann Arbor team members responded that we had no choice now about how to handle California. If we "give it away" in California, they said, other states would want the same flexibility.

At the outset of these deliberations, concern over precedent loomed large. An EPA action to approve test-and-repair stations in California surely would trigger similar requests from the nearly two dozen other states that needed to implement enhanced I/M. In addition, the deliberations took place at a time of growing concern over program integrity. As reported in the December 18, 1992, *Los Angeles Times*: "In the largest enforcement effort since the state Smog Check program began in 1984, more than 100 investigators swooped down on 24 Smog Check stations, a taxi fleet and a used car dealership Thursday and arrested 32 mechanics and station owners on felony fraud charges for allegedly trafficking in phony vehicle smog certificates." Thus, what remained unspoken during these early discussions, but was an ongoing concern, as illustrated by press accounts, was EPA's belief that test-and-repair programs were more subject to fraud than test-only systems.

As the date for California state senator Presley to introduce legislation approached, we spoke several times with senior CARB officials and with the senator's lead staff person, Carla Anderson. They delivered the same message: "If EPA opposes Gold Shield, the legislation is dead and California will not improve Smog Check."

EPA was forced to address the issue of what fraction of the total California program could be Gold Shield and still meet the performance standard. The agency was willing to approve a Gold Shield approach, but there was a catch: the state had to dramatically restrict the number of vehicles sent to Gold Shield stations for repairs and retests; otherwise, based on the deep (50 percent) discount EPA applied to emission reductions achieved by test-and-repair programs, California would not meet its mandated goals.

On January 14, 1993, I asked one of CARB's executive officers what the board envisioned for the Gold Shield program. CARB's view was that up to 40 percent of the cars tested might go to Gold Shield stations. The Ann Arbor staff was busy crunching numbers to estimate the small number of vehicles that could be tested by Gold Shield stations and still meet EPA's mandates. Clearly, though, 40 percent of the vehicle fleet was not the small number EPA had in mind.

On January 20 of that year, Bill Clinton was sworn in as president of the United States, with Al Gore as his vice president. In the weeks before the inauguration, the president-elect named key members of his cabinet and staff, including Carol

Browner as his choice to head EPA. Browner, prior to her appointment, was secretary of Florida's Department of Environmental Regulation; previously, she had been a member of then U.S. senator Al Gore's staff. Although Browner's selection was in place by inauguration day, it would take months for the Clinton administration to fill additional EPA posts.

By January 22, Gene Tierney's staff in Ann Arbor estimated that only 1 to 2 percent of all of California's cars could go through the Gold Shield program, receive no emission reduction credit, and still allow the state to meet its overall requirements. Ann Arbor's estimate was completely at odds with California's vision of having up to 40 percent of all cars tested at Gold Shield stations.

On Monday, January 25, just five days into the new Clinton administration, Senator Presley introduced Senate Bill (SB) 119, the state's legislative package to improve Smog Check, based on the IMRC report and the Gold Shield concept (Presley 1993). State legislative deadlines, the rush to meet the 1990 CAAA, and I/M implementation deadlines established by EPA all converged to propel the Smog Check issue forward at the beginning of 1993—right at the juncture when the federal government was transitioning to new leadership. EPA's new political leaders had barely arrived in their offices when they were faced with a wrenching decision: approve of California's Gold Shield approach and risk the domino effect of other states implementing test-and-repair programs, or disapprove of the approach and risk a political confrontation with California, the most important state in the nation regarding air pollution control, while sending a strong message to other states to adopt test-only systems.

By January 26, EPA's OMS leadership decided to unequivocally oppose Gold Shield, as well as California's preference to use the ASM dynamometer test instead of the EPA-designed IM240 test (Howekamp 1993). This key decision, taken less than a week into the Clinton presidency, set in motion the controversy that followed and wound up nearly unraveling the Clean Air Act. With this one action, EPA put itself on record that California's most politically viable Smog Check strategy was unacceptable to the federal government.

Once EPA headquarters decided to oppose Gold Shield, it became EPA Region 9's job to spearhead outreach to California stakeholders. During February and March, I worked with other EPA staff to meet periodically with constituent groups interested in an effective I/M program. Our task was to educate them about, as EPA viewed it, the merits of our I/M policy. We also needed to gain insight into the legislative process. As our outreach effort gathered momentum, we would be beset by calls to abandon our rigid policy stance. In addition, we would be lobbied to address—by either supporting or opposing—an array of alternative ideas.

In early February, Gene Tierney flew out from Ann Arbor, and together we met with legislative staff in Sacramento, including Carla Anderson in Senator Presley's office. Anderson advocated a Gold Shield program and chastised EPA: no state has implemented the EPA program, said Anderson, and there is no real-world data to support the agency's claims. Anderson made another point: we have a political problem with a group of scientists, she said. They were lobbying to use remote sensing devices (RSD) to augment or replace traditional Smog Checks. RSD used a roadside infrared light beam to measure pollution as cars drove by (see Figure 3-1).

If RSD proved effective, it might avoid having to require motorists to participate in traditional I/M programs such as Smog Check. At the time, however, RSD did not measure all pollutants covered by tailpipe tests and missed fuel evaporation problems. Also, as a car drove past an RSD unit, the unit captured just a brief snapshot of tailpipe emissions. EPA was unprepared to grant states substantial emission reduction credit for implementing RSD-based I/M. Anderson asked for EPA's help in putting down what she called the "RSD rebellion," warning that it threatened to derail Presley's efforts to improve Smog Check. As the California-EPA debate unfolded, RSD's proponents came to include, in addition to supportive scientists and law enforcement officials, several key legislators intrigued by the invention.

Much of our outreach effort revolved around the state legislative calendar. Many legislators, such as Senator Presley, introduced bills in January, and various committees held hearings on pending legislation throughout the spring and early summer. The legislature recessed in midsummer, then resumed session briefly before adjourning in September. With an eye toward a September end date, we in Region 9 anxiously tracked legislative developments, knowing that to meet clean air deadlines, California needed to adopt an EPA-approvable Smog Check improvement bill in the coming months.

In Sacramento, on February 16, the California Senate Transportation Committee held its first I/M hearing. The IMRC chose that date to release its fourth report to the legislature, which recommended Gold Shield and criticized the EPA model program contained in our November 5, 1992, I/M regulations (CA IMRC 1993). As the hearing began, State Senator Quentin Kopp, committee chairman, informed the packed room that the I/M issue would be "the most important issue the committee hears this legislative session."

From EPA's perspective, the hearing was a disaster. At one point, the service station industry showed a video of an evening news broadcast from Vancouver, British Columbia, which had recently implemented a test-only program. The newscast showed long lines and fuming motorists (not to mention cars) waiting to have their vehicles inspected. The hearing room was silent while listening to Canadian reporters interview exasperated motorists and watching inspections creep along at a glacial pace. The video had a huge impact. Dave Howekamp, briefing Dick Wilson later, called the hearing the "Great Train Wreck."

AUGUST 1990
VOLUME 40
NUMBER 8
(formerly JAPCA)

Journal of the
AIR & WASTE
Management Association

**Figure 3-1. An Air Quality Journal Features the Early use of RSD in Los Angeles. An Infrared Light Source is to the Left, a Detector to the Right (in front of barrier). A Van (at right) Collected the Data**

*Source:* August 1990 cover from the peer-reviewed *Journal of the Air & Waste Management Association.* Reprinted with permission

## A LINE IN THE SAND: MARCH–JUNE 1993

On March 3, we met with environmental group representatives from the Natural Resources Defense Council (NRDC) and Environmental Defense Fund (EDF). Instead of offering support, however, they aggressively questioned our scientific justification of the emission reduction benefits from test-only programs.

As our outreach progressed through mid-March, we were still struggling with staff in Ann Arbor to draw up scientifically based answers to basic questions. A critical issue was the technical justification for EPA's claim that test-only programs were twice as effective as test-and-repair. This issue, the 50 percent discount, would prove to be the most contentious in the months ahead. Ann Arbor staff said there was little primary evidence contrasting emission reductions across I/M programs. "The data are pretty slim," they warned those of us in the regional office. It would take until nearly the end of 1993 for Ann Arbor to publish a basis for the discount, and it would not be until 1995 that EPA provided more extensive documentation to rebut criticism of its regulations (U.S. EPA 1993b, 1995c).

Ann Arbor staff explained in public settings, as well as in private to those of us in the regional offices, that the MOBILE model—EPA's emissions simulation software—was the basis for the 50 percent discount.[1] MOBILE, however, was a black box to policymakers and even to many air quality professionals, as its embedded data and assumptions were not readily apparent.

On March 17, we met with CARB staff, who suggested we become more familiar with proposals by State Senator Newton Russell (Republican), from Southern California. Russell had introduced SB 1195 on behalf of the service station industry (Russell 1993). We did not realize it immediately, but the Russell bill would, for a time, represent the only major legislative alternative to Presley's Gold Shield proposal.

Several of us from Region 9 held a conference call on March 24 with scientists from EPA's Office of Research and Development (ORD) in North Carolina. Our goal was to get a more detailed understanding of RSD from Ken Knapp, a veteran EPA scientist who had spent years researching motor vehicle pollution. Ken told us some disturbing information; he said that RSD was not as bad as Ann Arbor staff portrayed it, and the IM240 program was not as good as they claimed. He had found that RSD could identify problem cars well; he had also run cars through IM240 tests, but they took longer than the Ann Arbor staff had said.

After the call, Ken sent us a memo he had written in October 1992 in response to other EPA questions about RSD, in which he said that 80 to 90 percent of the worst-polluting cars could be detected by setting up several RSDs to test vehicles under different conditions. He also sent us an article published the same month that claimed EPA opposed the new RSD technology in part because the agency had not originated the idea (Spencer 1992).

By April 1993, it was clear that the legislative process was boiling down to a choice between two proposals: SB 119, Senator Presley's bill to reshape Smog Check and launch a hybrid test-only and test-and-repair program with up to 40 percent of the vehicle fleet using Gold Shield stations, and SB 1195, Senator Russell's bill, sponsored by the gas station industry, to continue the existing test-and-repair program while requiring better enforcement. Neither proposal met EPA's requirements.

By the spring of that year, California legislators had not yet drafted legislation that could be approved by EPA, let alone be considered for adoption. Given California's political opposition to the agency's policy, EPA staff became increasingly nervous that they were about to lose the foundation to clean air. They believed that if California thwarted the agency's attempts to require rigorous automotive inspections, it would not only harm air pollution reduction efforts in the nation's worst-polluted state, but also set a trend that could sweep across the country. However, Congress, in the Clean Air Act, gave EPA a mandate to punish recalcitrant states by imposing sanctions. Congress also gave EPA some latitude to choose whether to proactively accelerate the imposition of sanctions, the most severe of which was the loss of federal highway funds. That spring, California was in the midst of a major recession. If sanctions took effect, the state stood to lose nearly $1 billion in annual highway funding. In the early 1990s, the California emission inspection business was a $500 million-per-year industry, employing thousands of technicians. Loss of either the existing industry or federal funds meant a substantial economic impact for a state in recession. A highway-funding sanctions threat was powerful but politically dangerous. Several EPA career staff remembered firsthand that the agency had successfully used sanctions a decade earlier, when California delayed creating the original Smog Check program; precedent, it seemed, suggested using sanctions again.

On April 13, EPA took the initiative to accelerate the use of sanctions by formally threatening to impose sanctions against California if the state failed to implement an I/M program that EPA could approve. EPA administrator Carol Browner and U.S. secretary of transportation Federico Pena cosigned a letter to California governor Pete Wilson that outlined the sanctions risk (the Browner–Pena letter). The letter attempted a positive tone but was nothing short of a veiled threat; either California delivered an acceptable Smog Check program, said Browner and Pena, or the state would lose its federal highway money (Browner and Pena 1993). "'The federal government is basically shaking the stick right now,' said James Lee, spokesman for the California Environmental Protection Agency. 'I don't think it's going to cause the Legislature to move any faster'" (*San Diego*

*Union-Tribune* 1993a). Although press reports questioned EPA's tactics, many of us at EPA Region 9 held out hope that the Browner–Pena letter would convince legislators they needed to create a program that met our goals.

April proved to be a pivotal month in the I/M debate. The sanctions threat threw the divide between the state and federal government positions into stark relief. Our continued meetings with state legislators yielded little legislative progress but much insight into the different views underpinning the disagreement.

In Sacramento on April 19, Dick Wilson, Dave Howekamp, and I met with various legislators to keep outreach alive following the sanctions threat. The meeting with Senator Quentin Kopp and his staff was contentious. Kopp, an Independent representing San Francisco, requested EPA documentation on the 50 percent discount, and his staff advocated on behalf of RSD.

We ended the day in Senator Tom Hayden's office. Hayden, a liberal Democrat and staunch environmentalist from Southern California, was impressed that EPA had the fortitude to send the Browner–Pena letter. Hayden said Senator Kopp was setting up what was being called a Blue Ribbon committee to evaluate Smog Check. The committee, said Hayden, included environmental, industry, and government representatives and was going to use the RAND Corporation to assess scientific information on I/M.

By April 29, I was back in Sacramento for more meetings, including with Senator Russell's chief of staff, who relayed a new view of the debate when she said the senator was incensed over what she called a states' rights issue, as well as the potential loss of jobs and consumer inconvenience. She said legislators, regardless of their political affiliation, were frustrated with federally mandated solutions to state problems.

I also visited Senator Art Torres's office and met with his aide. Torres, a Democrat, represented part of the Los Angeles region. Frankly, said the aide, if she spoke to five different people, she got five different proposals to solve the Smog Check problem. Without a scientific consensus to stand behind, she said, it made it difficult for the politicians to stick their necks out and support any one plan—especially if EPA was not going to support that plan.

I stopped by Senator Hayden's office and spoke with his aide, who said, "Tom will be for a program that is best for the environment. The problem is that nobody can agree on what the right solution is. The environmentalists are pushing RSD, CARB is pushing Gold Shield, EPA has a different position, and it's not clear what the right thing is environmentally."

My next stop was the office of State Senator Patrick Johnston, a Democrat representing California's agricultural central valley. His aide said, "We've tried centralized testing in LA before, and it didn't work." People visited the senator's office all the time and promised things just to get bills passed, she added, and 9 out of 10 of those promises never panned out. She noted that though EPA claimed people would not have long waits for a test-only inspection, no one knew what would really occur. Her comments proved to be some of the most prescient.

In the weeks that followed, the need to improve Smog Check remained a high-profile issue as the press reported efforts to reduce station fraud: "While the district attorney's office was investigating fraud among Smog Check operators last year . . . in October, they set up a remote sensing device at Los Angeles International Airport. . . . It was the first time such a technology has been used in law enforcement" (*Los Angeles Times* 1993a). Thus, just as EPA was trying to avoid having California rely too heavily on RSD to generate emission reductions, California officials were busy expanding its use in novel ways and generating press coverage that raised RSD's profile as an innovative new tool in the war against smog.

On May 14, Dave Howekamp informed me that the Ann Arbor staff was thinking about how a limited Gold Shield program could be structured, indicating that OMS was thinking of making a deal. Ann Arbor was searching for options. A month following the Browner–Pena letter, the sanctions threat had increased tension but had not yet helped resolve the dispute: "If [Governor] Wilson knuckles under to the federal Environmental Protection Agency, California motorists will get cleaner air but suffer the inconvenience of getting a smog test at one place and repairs elsewhere. If the EPA blinks first, drivers may wind up paying more but would still be able to get everything taken care of at a local gas station" (*San Francisco Chronicle* 1993).

A key problem, as evidenced by press coverage, was the sense that the state and EPA were locked into a winner-take-all battle. Lacking from the press reports and our outreach efforts, however, was any meaningful discussion of a compromise middle ground between the two sides.

As June began, the EPA regional staff began to network with the state assembly. On June 8, I traveled with my staff to Sacramento and met with Assemblyman Richard Katz, a Democrat representing the San Fernando Valley northwest of Los Angeles. Katz chaired the Assembly Transportation Committee but until now had kept a low profile on Smog Check. Later, Katz would become the central player in resolving the Smog Check dispute. This, however, was our first chance to engage him on the issue. As we entered Katz's office, the first thing I noticed was a poster that described how 10 percent of the cars caused 50 percent of pollutant emissions—a finding that had emerged, in part, from RSD-based studies. RSD advocates were using the findings to lambaste I/M: if only 10 percent of the vehicles were at fault, they argued, why inconvenience the other 90 percent of motorists by requiring their participation in Smog Check? The meeting with Katz did not go well.

On June 11, Gene Tierney and I traveled to the RAND Corporation's Santa Monica office to brief the Blue Ribbon committee created by Senator Kopp. During a tense six-hour meeting, we covered everything from the basis for the 50 percent discount to competing opinions about the efficacy of RSD. We were challenged on every point we attempted to make.

## KATZ BECOMES A MORE ACTIVE PLAYER: JULY 1993

On July 15, a staff member at CARB called and said Assemblyman Katz was getting more involved with the Smog Check issue. "He doesn't want to be the author, but he wants to broker a compromise."

During July, with the legislature in recess, Ann Arbor was still trying to figure out whether some version of a Gold Shield program could meet the EPA performance standard. As of late January, Ann Arbor had said that only 1 or 2 percent of the entire vehicle fleet could be tested at Gold Shield stations. The logic behind this, however, was premised on giving no credit to Gold Shield.

For a July 27 call with Dick Wilson and OMS staff, the Ann Arbor office had prepared analyses of whether some Gold Shield program might meet our performance standard. Phil Lorang explained the bottom line: fully half of failed vehicles could be allowed to go to Gold Shield by 1999, if California used IM240. If about 14 percent of vehicles failed Smog Check—a reasonable estimate based on experience—then about 7 percent of the fleet could be sent to test-and-repair stations. Dick's attitude toward the analysis was that it was a good exercise to go through in any case, just so we knew what the numbers looked like, but a Gold Shield program was still not approvable. After the call, Dave Howekamp said, "I'm not sure where Dick is headed with all this, but it doesn't leave me with a good feeling; how about you?"

By late July, our three-month old sanctions threat had not motivated an agreement. Part of the problem we faced was that many in California and elsewhere simply did not believe the Clinton administration had the political will to sanction California: "'They're talking big, but we have a hard time believing (Democrat) Bill Clinton would shut California down in the middle of a gubernatorial election,' said consultant Steve Schnaidt, whose [legislative] committee is reviewing two bills to revamp the smog-check program" (*San Diego Union-Tribune* 1993b).

Despite assurances from Washington, D.C. that EPA stood firm behind the sanctions threat, the press accounts provided those of us in the regional office with an independent view—one that sowed seeds of doubt about whether we would really ever sanction the state.

## POLICY COLLAPSE: AUGUST AND SEPTEMBER 1993

The legislature ended its summer recess, and on Tuesday, August 24, Senator Kopp's Transportation Committee held a hearing to receive the RAND report. The report represented an outside appraisal of the I/M issue coming from a highly regarded independent think tank. Its findings were devastating to EPA and triggered the eventual collapse of our national I/M policy.

The hearing was carefully orchestrated with the governor's staff, and with RAND, to support compromise legislation. RAND found no technical justification for EPA's preferred I/M program (Aroesty et al. 1993). Instead, there was general agreement on the need to strengthen the current program rather than scrap it for something else.

Jim Strock, California EPA (CalEPA) secretary and a member of Republican governor Pete Wilson's cabinet, testified at the hearing. Strock cited a speech Vice President Al Gore had given, in which Gore supported flexibility for state actions. Strock said the governor would have serious concerns about scrapping the existing program and moving to one that was centralized. From EPA's perspective, an especially ominous sign was that Republicans, Democrats, an Independent (Kopp), and the governor were all saying the same thing.

The day after Kopp's hearing, Dick contacted us from Washington, D.C., with news that, in hindsight, seems predictable but at the time was startling to those of us in Region 9: "We've been talking extensively with Senator Presley's staff over the past several days," said Dick, "and it looks like he has agreed to amend his bill." Dick later explained that the past several days had seen "a tremendous erosion of support for our position over at the White House." It had been clear for months that no political support existed in California for EPA's policy. The RAND findings made it appear that no technical support existed either. At this point, said Dick, we needed to do something dramatic to show flexibility and regain White House support for EPA.

On August 26, Administrator Browner sent a letter to Senator Presley approving a compromise (Browner 1993a). Dick called and walked Region 9 through the details: "The senator agreed to some significant changes that really strengthened SB 119, and once he agreed to those changes, he wrote to Browner and asked EPA to approve his bill if it passed the legislature. Browner agreed." At this point, everyone in Dave Howekamp's office looked at each other, as the realization that we had now officially caved in started to take hold. We moved on to discuss press issues, and Dick said Administrator Browner was going to give an exclusive interview to the *Los Angeles Times*. The strategy was to give *Times* reporter Melissa Healy a scoop on the story and hope she would write an article favorable toward EPA.

As Dick wrapped up his briefing, Dave asked for the key changes to Presley's legislation. "He expanded the bill to assure us that it meets our enforcement requirements," said Dick. "Future cars will only be able to go to Gold Shield if the program proves itself ... tell our contacts we haven't budged an inch on our performance standards." Once we understood the fine print behind the new Presley bill, it was apparent the folks in Ann Arbor were allowing about 7 percent of the vehicle fleet to be tested at test-and-repair stations: "Four months after sending a get-tough warning to state officials, the Clinton Administration on Thursday endorsed a compromise to overhaul California's Smog Check program. ... Browner said she endorsed the Presley plan because new amendments were added to crack down on the worst polluting cars, not because of political pressure" (*Los Angeles Times* 1993c). The press reported EPA's assertion that its position shift remained environmentally sound—but it failed to address an obvious question: if a compromise made environmental sense now, what had prevented EPA and the state from finding middle ground earlier?

EPA's action to approve the Gold Shield concept represented a modest shift in the calculus of estimating benefits from I/M programs. After all, Ann Arbor had moved from accepting about 1 or 2 percent of the vehicle fleet being sent to test-and-repair stations to about 7 percent. But the shift was a philosophical reversal. EPA went from adamantly opposing enhanced I/M programs that allowed test-and-repair to acknowledging that some combination of test-only and test-and-repair would be acceptable. After months of political pressure, EPA finally had to change course. The implications were not lost on other states struggling to maintain political support for EPA's test-only ideal. The Presley compromise marked a dramatic turning point in the months-long debate. Unfortunately for the agency, as discussed in the chapters ahead, EPA had just backed the wrong proposal.

## NOTE

[1] For a history of MOBILE, see NRC 2000; U.S. EPA 2004a.

CHAPTER 4

# STALEMATE: NEGOTIATIONS AND SANCTIONS

## SENATE TRANSPORTATION COMMITTEE HEARING, AUGUST 31, 1993

The Senate Transportation Committee heard the competing legislative proposals on August 31: Senator Presley's SB 119, revised to reflect the new compromise with EPA, and Senator Russell's SB 1195, revised to reflect the consensus among state officials following the RAND study. During the week leading up to the hearing, state political leaders had accelerated the pace of decisionmaking over Smog Check, and those of us in the regional office relied heavily on newspaper reports to supplement our understanding of the politics.

The political picture being reported the morning of the hearing looked bleak from EPA's viewpoint—the governor chose to support the Russell bill: "The long-running debate over how best to upgrade California's smog-check program got even murkier Monday when Gov. Pete Wilson disclosed that he won't support a compromise plan that has the blessing of the federal government. ... Wilson prefers SB 1195" (*Sacramento Bee* 1993). The morning newspapers dispelled all hope among the EPA regional staff that our compromise with Senator Presley would break the deadlock.

We were able to get and read copies of SB 1195 with only a few hours to spare. Russell's bill had just been rewritten and included compromise language acceptable to the governor and a bipartisan coalition of legislators. As we pored over the text, we discovered that the bill gutted the membership of the IMRC, replacing the committee's air quality professionals with political appointees, including representatives from the gas station industry. From EPA's perspective, the bill seemed a weak improvement of the existing Smog Check system and looked problematic for abandoning Smog Check's professionally based independent review panel.

As the hearing began, State Senator Quentin Kopp said SB 1195 had been amended to reflect the new consensus based on RAND's recommendations. "We'll take up the two bills before us, SB 119 and SB 1195, and in the tradition of the rules of this legislature, we will consider them in numerical order," said Kopp. He added, "SB 1195 is Senator Russell's bill; it is not the governor's bill. It represents bipartisan support achieved over the past four weeks of negotiation. . . ." Kopp continued, "The bill meets the Clean Air Act's requirements, which include flexibility for states."

Senator Presley spoke on behalf of SB 119, noting that his bill would create a system with private test-only contractors. "These will not be run by state employees; my bill does not extend the bureaucracy." Presley then emphasized five points regarding the bill:

- Smog Check was centralized, resulting in more convenience for motorists.
- Smog Check was cheaper because of mass-production-style testing.
- It created jobs. Service stations would lose the "sliver" of their business that did testing, but more problem cars would be caught, leading to more repair work.
- The program would have better quality.
- The bill had been accepted by EPA.

After senate committee debate, a vote was taken, and as Senator Presley had predicted to us before the hearing, SB 119 was defeated.

Kopp then moved the debate to SB 1195. Senator Russell discussed how RAND had discredited EPA's approach. SB 1195 "guarantees compliance with the federal standards," said Russell. "If the standards are not met, then additional controls will be implemented."

During the debate, Senator Hayden raised questions about SB 1195:

**Hayden**: "So your bill guarantees it will meet federal standards?"

**Russell**: "Yes."

**Hayden**: "And it does this by checking to see if it has met the federal standards, and if the standards haven't been met, then further controls will be implemented?"

**Russell**: "Yes, that's correct."

**Hayden**: "Now, how will the determination be made as to whether your bill has been effective or not?"

**Russell**: "The I/M Review Committee will make that determination."

**Hayden**: "You mean to tell me that you will be willing to accept the I/M Review Committee's analysis . . . "

**Kopp**: "Well, now, let me try and answer that question . . . "

**Hayden**: "I haven't gotten to ask my question, Senator Kopp. How can you answer my question? You don't know what my question is. I'd like to finish asking my question. No, no, that's OK. Senator Kopp, why don't you just tell me what my question was. That would be better."

[laughter from the audience]

**Hayden**: [back to questioning Russell] "You mean to tell me that you will be willing to accept the I/M Review Committee's analysis of whether your bill has been effective or not? Isn't this the same I/M Review Committee that said the current program is a failure and recommended to us that we adopt Senator Presley's bill? What are your comments, then, to the current I/M Review Committee analysis?"

**Kopp**: "The current I/M Review Committee was never officially constituted."

**Hayden**: "The current I/M committee says that the program you're proposing will not get the reductions EPA wants..."

**Russell**: "Well, it's not the same committee."

**Hayden**: "It's not the same? The bill says it's the I/M Review Committee—isn't that the same committee staffed by professionals from the air quality community?"

**Kopp**: "Well, the bill changes the composition of the committee."

**Hayden**: "Ahhhhh!!! The bill changes the composition of the I/M Review Committee. *I seeeeee.*" [extended laughter from the audience] "So who gets to sit on the committee now?"

Russell then detailed how his bill allowed political appointees on the committee. Although EPA staff realized before the hearing that SB 119's defeat and SB 1195's passage was assured, we watched with amusement as Hayden, the committee's most outspoken environmental advocate, dissected the bill's reworking of the IMRC. At the time, though—in the "us" versus "them" atmosphere that prevailed—I doubt whether any of us from EPA gave enough credit to the idea that by changing the IMRC from being composed entirely of air quality professionals to being composed of a mix of representatives from different stakeholder groups, Russell's changes would expand public participation in, and improve oversight of, the Smog Check program.

After the committee debate, Kopp said that if SB 1195 was approved, it would probably be amended into another bill to be considered in a conference committee. The legislators then voted. As expected, the Russell bill passed.

The bill EPA supported had failed, and something altogether different had passed. The state was headed toward sanctions. Worse yet, the time available to find an EPA-acceptable legislative solution was passing quickly: the last day of the legislative session was scheduled for Friday, September 10, barely a week and a half away.

## TOWARD A NEW STRATEGY: SEPTEMBER—NOVEMBER 1993

On September 1, Dick's deputy called and said that EPA's political appointees and career staff had met that morning with Tom Epstein, a member of the White House staff, to brainstorm what to do following passage of the Russell bill. Also in the group was Mary Nichols, the new appointee to serve as EPA's assistant administrator for air. Mary was the former chairperson of the California Air Resources Board and a well-qualified Californian. Upon her arrival at EPA, nearly eight months into the Clinton administration, she was handed Smog Check to resolve.

The group said EPA had not yet made its case about jobs, cost, convenience, and emission reductions attributable to the EPA program; they advised focusing more energy on those messages. White House staffer Epstein recommended that EPA administrator Browner herself hold a press conference the next day to address the issue.

The group also discussed a request from Assemblyman Richard Katz to review yet another piece of Smog Check legislation, SB 629 (Russell et al. 1994). Katz planned to use this bill to carry any compromise that might emerge with EPA, and he wanted the agency to explain what would make the bill acceptable. SB 629 ultimately became Smog Check's main legislative vehicle, absorbing the ideas approved by Senator Kopp's committee as SB 1195.

On Thursday, September 2, the EPA administrator held an I/M press conference with reporters from every major California newspaper. Referring to the upcoming adjournment of the legislature, Browner informed California's press corps, "In eight days we will enter the sanctions process if we don't get legislation."

On September 8, I testified about the possibility of upcoming sanctions to the California Transportation Commission in Sacramento. While there, I learned that the assembly had approved SB 629 on a lopsided vote (65 to 9), sending the bill to the state senate.

Early on Friday, September 10, the last day of the state legislative session, in a last-minute attempt to forestall sanctions, Browner agreed to delay them in exchange for the state not passing SB 629. In response, the senate president pro tempore David Roberti killed SB 629 by holding it in committee.

Although the sanctions process was delayed, legally the state owed EPA an I/M plan by November 15. Browner, in her September 10 letter to Roberti, said, "I am committed to withhold the proposal of any sanctions until the November 15, 1993 deadline, and to delay the imposition of sanctions for a period of time sufficient for my staff and the legislative leadership to craft an acceptable bill which can be enacted soon after the legislature reconvenes in January, 1994" (Browner 1993b). Although EPA continued on a path to propose sanctions on November 15, we were prepared to halt that process if a compromise could be developed in time.

The long legislative road had ended in a stalemate. Region 9 had instructions to be prepared to make good on the sanctions threat on November 15. But the path to November 15 was poorly marked, at best. In the space of two weeks, EPA had twice shifted positions. First, on August 26, Browner said that Gold Shield would be acceptable (Browner 1993a). Then, on September 10, Browner delayed sanctions in return for blocking SB 629.

As those of us in Region 9 digested what had taken place, the state's key legislative players worked quickly to keep the issue moving. In a letter to Browner cosigned late on Friday, September 10, legislators Dan Boatwright, Willie Brown, Richard Katz, Quentin Kopp, and Newton Russell all committed to negotiate a solution. They enclosed a copy of SB 629, asked that EPA review the bill, and expressed hope that discussions would begin soon (Brown et al. 1993).

Once the legislature adjourned, and the debate dragged on, journalists took more time to delve into the technical side of the dispute: "'My conclusion after digging through all these reports is that a centralized (state-run) system, while a possible benefit, is not the panacea EPA makes it out to be,' said Janet Hathaway, an attorney for the NRDC, an environmental group" (*San Jose Mercury News* 1993a). Press reports such as this one about the technical shortcomings of our position, when paired with reports of our policy shifts, further eroded our ability to defend EPA positions and maintain the threat of sanctions.

On September 17, Dave Howekamp was in Washington, D.C., where he met with White House staff member Tom Epstein, Mary Nichols, and others. EPA had made a clear statement of intent to sanction California by mid-November, and the White House, in the form of Epstein, now took a more active role. Dave reported back to Region 9 staff that Mary did not think sanctions would make California implement EPA's program. She said there was no buy-in to EPA's technical solutions, and that the agency would have to negotiate a resolution.

On September 21, we had a follow-up call with Ann Arbor, Dick, Mary, and Epstein. Epstein pushed for an aggressive press outreach campaign, saying EPA had not done enough to educate reporters.

As the White House and EPA's headquarters staff worked to develop a new Smog Check strategy, a fresh perspective joined Region 9. After some nine months in office, the Clinton administration finally appointed someone to head EPA's San Francisco office: Felicia Marcus was sworn in on October 4, 1993. The Smog Check debate was one of the first problems she faced.

Felicia's arrival marked an important turning point for Region 9's involvement in Smog Check. She brought substantial environmental expertise to the debate, from an entirely new perspective—what it was like to be on the receiving end of federal regulations— and, in retrospect, demonstrated the unique contributions political appointees can make when collaborating with career staff. Before joining EPA, she ran the Los Angeles Public Works Department. She had a reputation as an effective negotiator, able to bring unlikely allies to agreement. In fact, she ended up running the Public Works Department even though she had previously been outside the organization suing them as a public advocate. Felicia also had served on the governing board of the Coalition for Clean Air, an environmental organization, and had represented environmental organizations on issues that came before the Los Angeles area's regional air pollution control agency. Her community advocacy work brought grassroots air quality expertise that enhanced EPA's credibility when it came time to broker a deal with the state.

Also on October 4, while Region 9 staff developed the paperwork to sanction California if required, President Clinton visited the San Francisco area, promising highway funds. Clinton committed the federal government to give $318 million to help rebuild the Cypress Structure, a double-deck freeway in San Francisco's East Bay that had caught the nation's horrified attention when it collapsed during the 1989 Loma Prieta earthquake. The juxtaposition of our staff work against the president's message could not have been more stark; at that point, few of us among the Region 9 career staff believed the administration would ever sanction California.

On October 13, we briefed Felicia Marcus on the I/M issue. She asked why EPA couldn't just modify the current program, as Katz had suggested. I outlined that

California had already tried to improve its program, but it did not work; that a California Smog Check was already the most expensive test in the country; and that Katz's proposal would only make it more expensive. This was EPA's standard thinking about Smog Check. With her fresh perspective, though, Felicia was asking something more fundamental: why oppose letting California proceed as it thought best? Her question, though disarmingly simple, suggested the mindset change needed to reach a deal.

As staff prepared sanctions paperwork, press reports indicated that Browner was having difficulty securing White House support for EPA: "Though the EPA is chronically underfunded, Browner was unable to prevent her first budget from being cut. The White House delayed for months the nominations of her key political staff" (*Rolling Stone* 1993).

The more the press speculated about the limitations of Browner's influence, the more convinced the Region 9 I/M team became that our sanctions efforts were wasted time. In addition to the reported lack of White House support for EPA, a gap emerged between the Clinton administration's environmental policy actions during its first months in office and public expectations about how the actions of the new administration would differ from those of its predecessor: "After 10 months in office, the Clinton Administration has done next to nothing to improve environmental policy. . . . Instead Clinton's environmental agenda seems to have fallen off the radar screen" (*Los Angeles Times* 1993b). The reported gap between Clinton administration expectations and actions contributed to a shift in expectations among agency staff and to a growing sense of uncertainty within Region 9 over the eventual outcome for I/M.

As November began, Mary said she expected Browner to sign official paperwork on November 16 that would begin to implement sanctions. While discussing how to reach a compromise, Mary instructed us that EPA should accept *any* alternative that was environmentally sound.

On November 2, Mary included Epstein from the White House in a conference call to keep him apprised of our plan to move forward with sanctions. The White House wanted to avoid having Browner sign the sanctions package on November 16, as that date would fall close to an important upcoming congressional vote. Epstein was also concerned because the president planned to visit California at the end of that week. As an alternative, he suggested announcing sanctions during the week of the Thanksgiving holiday. "Great," quipped Dave Howekamp, "two turkeys in one week." Our dual charge was now to prepare to start the sanctions process but also to find a compromise that could work.

On November 9, Dave Howekamp and I ventured back to Sacramento and assembled in a conference room with staff representing each of the key legislators. The hostility was palpable, with comments about how these discussions should have taken place months earlier. During the discussion, the state's delegation was skeptical about working out an agreement.

The conversation focused on an idea proposed by Assemblyman Katz: what he called a "parallel path." The concept was to run two programs simultaneously—EPA's approach and California's approach—and then see whether the state's method proved acceptable over time. Dave and I viewed the parallel path as impractical. In our view, there was no feasible way the state could create two different programs, implement them for sufficient time to gather meaningful data, decide which program to pursue, and fully implement a final choice, all within the brief time allowed under the Clean Air Act and EPA regulations. Dave and I were careful to describe what would need to be accomplished to make the parallel path work without directly shooting the proposal down. Finally, the others also recognized that federal deadlines made the parallel path impossible to complete. When the meeting ended, we had little to show for the effort.

It would have been easier for EPA staff to absorb the problems we were having with I/M if we were succeeding on other fronts to achieve clean air. However, EPA's clean air efforts were bogging down: "Three years after Congress rewrote the Clean Air Act, the Federal Government and the states are consistently behind on many of the law's demanding timetables. ... 'EPA must improve its performance ...' said a report written by Republican and Democratic staff members of the Senate Environment Committee" (*New York Times* 1993). Our delayed progress on Smog Check was symptomatic of the larger delays experienced by EPA and the states as they struggled with the 1990 CAAA's ambitious requirements. Many of the act's deadlines were proving impossible to meet.

By late November, we had made no progress in resolving the dispute. Consequently, as most Americans prepared to celebrate the Thanksgiving holiday, the EPA staff was preparing to announce that we had started the process of withholding nearly a billion dollars from California. On November 22, 1993, EPA staff began calling interested stakeholders external to the agency to brief them about the upcoming sanctions announcement. This was the start of the sanctions communications strategy.

The communications strategy is essential to any important government announcement. A main function is to brief key stakeholders so they are not surprised when reading or hearing about a policy via the media. For example, our I/M communications strategy included preannouncement contacts with the politicians who had taken an interest in the I/M issue, such as Assemblyman Katz. In addition, a key communications goal involved working with the press in advance of the sanctions announcement to make sure they understood the decision being made. This also gave the agency a chance to make its case to the media. As an example of our media outreach plan, our communications strategy included briefing editorial board writers of the state's major newspapers. The I/M sanctions announcement was considered so important that Mary Nichols and Felicia Marcus were called on to make many of the press contacts themselves, in addition to using our normal press liaisons.

Tuesday, November 23, was the critical preannouncement communications day. Mary had flown out from Washington, D.C., the night before, and on Tuesday morning we all assembled in Felicia's office to review the day's schedule. It was now just one day before the sanctions announcement, and all of the important communications work had to be completed over the next 8 to 10 hours. As we ran through our notes, Mary mentioned that as we were speaking, Richard Katz was in Washington, D.C., meeting with Carol Browner.

A little after 4:00 p.m., Dave Howekamp came to my desk. Dave was an unflappable personality, a leader who remained calm in almost any storm. But I could tell something was wrong. "Come to my office," he said. "I don't have a good feeling about this," I joked. "You shouldn't," he replied. Bill Glenn and Virginia Donahue, our press liaisons, also came to Dave's office, and then he dropped the bombshell: "The announcement's off." Dave explained that the administrator and several of her staff had met with Katz in the morning to talk about Smog Check. But it seems that nobody at the meeting told Katz we were proposing sanctions the next day. In the meantime, Region 9 had briefed Katz's California staff as part of our communications strategy. Dave told us that apparently Katz was incredibly angry when he heard the news secondhand from his staff. Katz then spoke with the administrator, who apologized for not telling him about the announcement and agreed to delay the sanctions process.

Virginia immediately left the meeting to halt the further release of our press packets, and then we met with Mary and Felicia. Now the task became figuring out how to portray the situation and what to do in the coming days. After some brainstorming, the staff agreed that the reason for delaying the announcement was to explore the "significant developments" that had taken place in the last several days, alluding to Katz's meeting with Browner. The next morning, the day before Thanksgiving, we held a strategy session with Mary, Felicia, and others from Washington, D.C. Mary agreed that she would complete the difficult task of calling back newspaper editorial board staff and explaining EPA's about-face.

During a phone call with the D.C. contingent, we discussed how to develop a substantive negotiation strategy, as we were now touting the "significant developments" that allowed us to postpone sanctions. Mary offered ideas about how we might agree to a pilot study to determine how much credit to give to the Katz plan. Some EPA staff commented on the limited flexibility we had because of the 50 percent discount for test-and-repair programs. Mary replied that she had read our regulations and had defended, and would continue to defend, the

50 percent discount. "But," said Mary, "anyone who has looked at this information for about an hour sees that it is no more than a guess on our part about how effective these programs are. I think we may have to consider showing more flexibility here, because it's something that may be warranted given the lack of data."

Later, Nancy Sutley, Browner's air quality aide, told us the administrator planned to call key California legislators on the Monday following Thanksgiving and wanted to be able to recommend a negotiation process. Browner's idea was to include all the legislators in the process, but to focus on working with Assemblyman Katz to come up with substantive proposals. She wanted negotiations to take place sometime over the next two to three weeks, with Felicia in the lead for EPA.

EPA's Washington, D.C., media office issued a statement later in the morning that, as a result of the "recent progress" in the negotiations with California, the agency had "decided to call a halt temporarily to the process of imposing discretionary sanctions" (Nichols 1993). It did not take long for word of EPA's abrupt decision to have an impact across the country. Within hours, for example, the deputy secretary for the state of Louisiana's Department of Environmental Quality halted efforts to have the state legislature approve a centralized, test-only I/M program (Kucharski 1993).

EPA's statement led to press reports of an imminent compromise; however, press accounts also focused on the fact that the agency once again had shifted direction: "The EPA had scheduled a news conference in San Francisco Wednesday morning to start the clock ticking on withholding highway funds. Then, in a surprise move that one high-ranking state environmental official said 'left us flabbergasted,' the EPA reversed course" (*Wall Street Journal* 1993). The press accounts helped further reinforce among Region 9 staff the apparent political infeasibility of implementing sanctions against the largest state in the nation.

## OPTION E: NOVEMBER 1993–JANUARY 1994

Staff from Ann Arbor worked over the Thanksgiving weekend, and by Monday, November 29, EPA had some new options on the table. After hours of computer work, and what must have involved some substantial opinion shifting, Ann Arbor created several new I/M options that would have been unacceptable just days before. Even though the agency's original test-only policy ended with the August compromise with Senator Presley, up until this point EPA staff had tried to retain as much of the test-only plan as possible. However, after sustained political pressure for nearly a year and repeated policy reversals on sanctions, Ann Arbor finally had abandoned its strategy to mandate that virtually all of the state's fleet be inspected at test-only facilities.

The most important of Ann Arbor's new options was that cars that were no more than five years old could go to test-and-repair stations. In California, that meant roughly 40 percent of the vehicle fleet. This new option was in stark contrast to the compromise struck with Senator Presley, in which EPA had said 7 percent of cars could go to test-and-repair facilities. It was a seismic shift from where the agency had been in January, when Ann Arbor said only 1 to 2 percent of the fleet could go to test-and-repair. Ironically, we had now reached nearly the same position the California Air Resources Board had advocated nearly a year earlier, when CARB's senior management envisioned up to 40 percent of the fleet being allowed to go to Gold Shield stations. The option to allow 40 percent of vehicles to go through test-and-repair facilities was Option E in a list of eight options (A through H) prepared over the Thanksgiving holiday.

On Friday, December 3, Felicia officially shared the options paper with Katz, and by formally releasing the options, we committed ourselves to providing these options not just for California, but for other states as well. It was a dramatic change in national policy (U.S. EPA 1993c).

## The Real Negotiations Finally Begin

The week of December 6, 1993, marked a turning point in the now year-long debate. For the first time, EPA met with representatives from the state legislature with the intent not of educating them about the agency's requirements, but of negotiating a resolution to the dispute. By Monday morning, Felicia reported that Katz had looked over the options paper and was ready for discussions to begin. The first negotiating session was set for Wednesday.

As we prepared for negotiations, Felicia said Katz did not understand how EPA's computer model, MOBILE, estimated a 50 percent discount for test-and-repair programs, and he also wanted to understand what credit we gave to RSD. Felicia emphasized that at the negotiations, we needed to be able to share in plain English how EPA credited I/M programs. Felicia then set our new tone: "We need to be more flexible here. California is trying to come up with an alternative model; we should recognize that they're working hard on this and give them more credit for what they're trying to do."

## The First Negotiating Session: December 8, 1993

On December 8, Dave Howekamp, Felicia, and I met with Katz's staff and others. A new EPA staff member, Cecilia Estolano, also joined us. Mary had just hired

Cecilia as one of her personal aides, and the choice was fortuitous. Cecilia not only brought excellent analytical skills, but also had the perfect employment history: her old boss was Assemblyman Richard Katz.

As people filed into the conference room, we took seats around a large, rectangular table. John Stevens from Katz's office sat at the head of the table and chaired the meeting. Present were staff representing all of the major state players, including the senate president pro tempore David Roberti; Senators Presley, Russell, and Kopp; Assembly Speaker Willie Brown; and representatives from the California Environmental Protection Agency (CalEPA) and the Bureau of Automotive Repair (BAR), the agency that ran Smog Check.

John Stevens began the first negotiation session by saying, "We are not here to cut a deal—we are here to explore issues." Stevens, who would lead many of the upcoming discussions with great skill, started by laying out the key goal. He hoped we could devise an objective evaluation process to see if the test-and-repair program outlined in SB 629 was effective enough to meet EPA requirements.

Felicia then gave some introductory remarks: "I hope the focus of these meetings is to have a conversation ... to try for the first time to sit down and have a normal dialogue about these issues." She said the key question was whether there was room for movement toward a solution.

During the meeting, Felicia signaled EPA's new desire to bend over backward and work with the state. Even though Katz had privately let Felicia know earlier in the week that Option E looked good, during the meeting John Stevens and CalEPA deputy secretary Mike Kahoe focused on the parallel path, or dual-track, program—the idea of implementing two different approaches and then comparing the results to see what worked.

Thus ended the first of what ultimately would be a dozen face-to-face policy-level negotiating sessions in Sacramento; technical negotiations to resolve scientific disagreements; and supplemental meetings that involved state legislators, the governor's office, and EPA's most senior political appointees, including Administrator Carol Browner. Felicia laid the groundwork for many of these sessions by traveling to Sacramento alone or with Cecilia and spending hours rebuilding relationships with state officials. These intensive discussions lasted from early December 1993 through March 1994. They were rocked by collapsed talks, EPA sanctions, and a natural disaster. Yet they ultimately bore fruit, resulting in a novel program design and a negotiated agreement that had national repercussions.

## The Second Negotiating Session: December 16, 1993

On December 16, we held our second negotiating session in Sacramento. In contrast to the first session, however, this one proved more frustrating, as the group reached an impasse over technical issues. The largest problem was that the California contingent did not believe that EPA's 50 percent discounting of test-and-repair programs was either fair or technically justified.

We closed by setting a time frame for follow-up negotiations. With the holidays approaching, we could not reconvene the group before the new year. John Stevens, from Katz's staff, said, "We'd better try and resolve this thing before the legislature gets going again." We left with a loose understanding that we would aim for another staff-level meeting on Tuesday, January 4, with a principals' meeting the following Tuesday, January 11. This second date was to prove critical. State negotiators believed they had a good-faith representation from EPA that we would remain engaged in negotiations until the January 11 principals' meeting—a commitment we were unable to keep.

## Technical Negotiations

On December 20, 1993, the first technical negotiations were held. During the telephone call, EPA and CARB staff parried over mind-numbing technical minutiae, and we accomplished little.

Later that afternoon, we heard alarming news from Felicia: Administrator Browner remained under pressure to move forward with sanctions because of threats from other states to undo test-only I/M programs adopted in recent months. Browner had wanted to make the sanctions announcement within the week, but Felicia and Mary persuaded her to wait until January 5. This was a huge problem for those of us negotiating with the state, as during our last negotiation, we had agreed on the January 11 date to cement a deal.

Unfortunately for all of us, the California I/M discussion was taking place on two different tracks. One track was in California and involved the policy and technical negotiations that we were orchestrating from San Francisco. A second track was in Washington, D.C., where high-level political discussions were taking place among Mary Nichols, Carol Browner, and the White House. The tracks had just diverged. In one of the strange twists of the Smog Check story, EPA staff on opposite ends of the country were working toward different goals. Region 9 was motivated to compromise with California to secure an improved Smog Check program. In contrast, the D.C. contingent was motivated to sanction California if a deal was not imminent. Their logic was that sanctions would demonstrate to other states that EPA was serious about the enhanced I/M mandate, and they hoped this would maintain momentum in those states to adopt and implement test-only programs.

### Scheduling the Sanctions Deadline

On December 22, Mary told us that sanctions would probably not move forward until the second week of January, because Browner would be in California during the week of January 10, and EPA did not want to propose sanctions until after she had left the state. Mary said the White House would contact California's legislative leaders with a deadline for concluding a deal. It was stressed that we in Region 9 were *not to mention* the impending threat of sanctions, nor the deadline, to any of the people with whom we were negotiating. After EPA's recent policy reversals, there was tremendous sensitivity at the agency about threatening sanctions but not following through. Any talk of sanctions now had to come from the highest levels.

As the days passed, the sanctions announcement date kept changing. Legislative deadlines were fast approaching in other states. EPA wanted to have either a California deal consummated or sanctions announced before the other states lost their window to approve I/M legislation. As the press was reporting: "California's tussle with the federal government over revamping the Smog Check program is sparking revolts in other states. ... From Nevada to Pennsylvania, quiet grumbling has given way to loud complaints ... since federal officials delayed penalizing California" (*San Jose Mercury News* 1993b).

Though the press reports captured the overall conundrum EPA faced, they rarely shed light on the strange dynamic that existed within the agency. The more we in Region 9 sought to find a workable compromise with California, the more risk we created for our colleagues in other regions. EPA regional offices that had already helped their states secure test-only programs had nothing to gain and everything to lose from a brokered deal in California. The national fallout from the debate loomed large, signaling the mounting—and conflicting—pressure on EPA.

## SNATCHING DEFEAT FROM THE JAWS OF VICTORY: JANUARY 1994

With the New Year came a burst of negotiation work. On Monday, January 3, 1994, we briefed Administrator Browner about our plan to spend Tuesday through Thursday in Sacramento, trying to work out a deal. Felicia explained to Browner that at Tuesday's negotiations, we expected CalEPA to present us with a dual-track option, which meant the state would simultaneously implement EPA's and the state's preferred programs. Sometime in 1994 or 1995, after data became available, the state wanted us to decide whether its program could meet our performance standard. At tomorrow's meeting, counseled Felicia, we planned to emphasize the need for other options and question whether the state's schedule to implement the dual-track program was reasonable. "I'm not sure if we are having a meeting of the principals this week," she closed.

As we talked, we learned that the White House had not yet communicated to California's legislators the deadline to reach a deal or risk having the sanctions process begin.

It was abundantly clear the strain Carol Browner, Mary Nichols, and Dick Wilson were under. Continued uncertainty over the I/M outcome in California was causing other states to slow or halt their legislative efforts to adopt test-only programs. And a cause for special concern was that in Illinois and Indiana, we might soon lose the legislative window entirely. The Illinois legislative calendar had the most important near-term deadlines. Illinois would be in a brief legislative session the following week, lasting only Wednesday and Thursday, January 12–13.[1] After Thursday, the Illinois legislature would not reconvene until March. EPA's Washington, D.C., staff felt strongly that because of Illinois, the agency needed to make the California sanctions announcement on Monday, January 10. That would allow press reports to come out on Tuesday, in time for Illinois legislators to see the news before they began their two-day session on Wednesday. As one senior EPA career official noted years later, "It would be impossible to overstate the pressure EPA was under from states to hold the line on the test-only requirement."

## The Third Negotiating Session: January 4, 1994, the Reverse-Trigger Breakthrough

On January 4, Dave, Felicia, and I headed to Sacramento. The negotiation participants that morning included those who had attended the previous sessions, as well as Assemblyman Katz and Senator Russell.

Katz offered opening comments, emphasizing that time was short and the governor's office wanted the issue resolved within the next two weeks. He said we needed to wind up "somewhere between SB 629 and Option E," and he acknowledged the significant compromises that had occurred on both sides. The assemblyman then said he had some ideas for everyone to consider. What happened next was a tremendous breakthrough: Katz said he was prepared to support a program in which the state would commit up front to meeting the requirements of Option E, and then back off from that commitment depending on the effectiveness of the state's program. He called this a "reverse trigger" because it was the reverse of SB 629, in which the state committed only to its plan, and then switched to EPA's approach if the state's plan did not work. The quid pro quo to Katz's offer was that EPA had to work with the state on a pilot study to evaluate SB

629's effectiveness. The state would have to be allowed to replace Option E with whatever proved effective from SB 629. Katz had offered us a major concession, and if we could get everyone to agree, we had the seeds of a deal.

Katz again emphasized timing and said he was concerned we would lose negotiating momentum if we waited two to three more months. He then said we should have a principals' meeting next week, on January 11. The assemblyman closed with a warning: "If EPA sanctions the state, then there will be a negative reaction from the legislature."

We left after identifying the critical issues facing both parties. The state had two main concerns:

- How do we evaluate SB 629?
- Would EPA buy into the results?

For EPA, the following were key concerns:

- Would the state's pilot study have firm deadlines?
- Would the state meet EPA's phase-in goal of sending 15 percent of vehicles to test-only Smog Checks starting January 1, 1995?
- Would test-only vehicles be tested using IM240, which EPA wanted, or the less expensive ASM test, which the state preferred?
- Who was buying into this agreement? Did we in fact have a deal with all the key legislators and the governor?

It was 8:00 p.m. in Washington, D.C., when Felicia reached Browner's chief of staff Mike Vandenbergh and air quality aide Nancy Sutley, Mary Nichols, and the administrator herself to explain what had happened. Felicia asked what we needed from California to avoid sanctions and say we had a deal. Browner told her that in addition to having a substantive agreement we could support, we needed a commitment to the deal from California's legislative leaders.

What was increasingly obvious by the end of Tuesday, January 4, 1994, was that the road to a Smog Check deal traveled through the office of Assemblyman Richard Katz. With his reverse-trigger proposal, Katz cemented his position as EPA's most important political ally in California.

## The Fourth Negotiating Session: January 5, 1994

The next morning, Wednesday, January 5, Nancy Seidman, my counterpart in EPA's Boston office, left me an ecstatic voice mail message: "Doug, I just wanted to let you know that we got our I/M bill in Massachusetts. The legislature passed it

20 minutes before they recessed! Good luck in California!" It was a rare piece of good news that despite the problems in California, at least some states appeared willing to continue forward under EPA's plan. Soon after, Dave, Felicia, and I headed to Sacramento.

As we began our negotiating session with Katz, Russell, and the by now all-too-familiar participants, the discussion tone was tense, with nerves frayed and anger near the surface. It would have been easy for an outsider to assume we were far apart. Yet we were making substantial progress. Katz had shifted the entire debate.

As the meeting neared its end, it became clear the whole deal was contingent on creating an evaluation protocol everyone would agree to. The months of posturing, antagonistic press stories, and threats had taken their toll; though we were nearing an agreement, everyone felt we needed to write down the details to avoid later misinterpretation. We discussed creating a memorandum of agreement (MOA) binding EPA and the state to the results of a pilot study. Katz and Stevens then both mentioned the need to wrap up the negotiations "in time for the Tuesday, January 11, principals' meeting."

After the meeting, we called Washington. Mary dropped a bombshell: the sanctions announcement could come as early as Friday, January 7—just two days away. Mary explained that the vice president would soon travel to California, and the White House wanted to place as much time as possible between his trip and the sanctions announcement. This was the first any of us, including Felicia, had heard of this development. And the legislators with whom we were negotiating had no clue about our internal deadline—the White House had never communicated it to them.

## Sanctions

Thursday, January 6, was all-or-nothing day. Either we reached an agreement or within 24 hours headquarters would make its sanctions announcement. The day began with a rapid-fire series of phone calls that took place in a desperate, pressured atmosphere.

At 7:00 a.m., I was on the phone to Ann Arbor. Dick outlined his concerns about California's pilot study and listed questions we needed to have California answer.

At 8:00 a.m., Dave and I had a call with EPA in D.C. to discuss the mechanics of how the sanctions package would be announced.

At 9:00 a.m., Dave and I made another phone call to Ann Arbor. We discussed in more detail how to put together a pilot study to test California and EPA's programs against one another.

At 9:30 a.m., Dave and I orchestrated a conference call involving Ann Arbor and technical staff from CARB and BAR in California. The sense of urgency was apparent. As we closed the call, Dick asked how long California would need to finish the pilot study and share results with EPA. The state indicated that the study results might be available by the fall, a mere 9 to 10 months away. Dick thought that time frame would work. After the call, we had another EPA-only conversation, and Dick sounded very positive about the ability to put together a deal. We just needed to work out the details.

As the afternoon wore on, it became apparent that it was going to be impossible to tie the loose ends together. We were extremely close, but the people we were negotiating with had no idea the deadline had been moved to that very day. The White House had never let them know January 11 was too late. As it was Thursday, legislators were already leaving the capital for their districts. Katz had left for Los Angeles in the early afternoon.

Dave and I worked feverishly in the late afternoon to edit the state's proposal for the pilot study, and Dick was participating by phone. But it was getting late. While we talked to Dick, Felicia called Mary. Around 4:00 p.m., Felicia returned, looking defeated, and told Dave and me that we were going ahead with sanctions. It was her job to call Katz and let him know. The turmoil showed on Felicia's face. She sat alone in a side office, composing herself for the phone call. Dave and I waited quietly in the next room. When she finally reached Katz and told him we were moving forward with sanctions, he became incensed. They spoke for 20 minutes or so, during which the conversation mostly consisted of Katz—justifiably—venting his frustration.

The next morning, Friday, January 7, 1994, Carol Browner signed the official sanctions *Federal Register* notice, and EPA announced that it had started the sanctions process against California. To maintain I/M policy consistency across the country, EPA simultaneously started the sanctions process against Illinois and Indiana as well (U.S. EPA 1994c).

The *San Francisco Chronicle* reported the next day that "after a brief truce, the U.S. Environmental Protection Agency declared war yesterday on California's Smog Check program, taking the first step toward imposing sanctions. ... The move caught state lawmakers involved in negotiations with the EPA by surprise" (1994a). An irate Katz was quoted in the *Los Angeles Times* (1994a) as saying, "EPA is snatching defeat from the jaws of victory."

Although the press tried their best to fathom what had taken place between EPA and the state, none of the newspaper accounts discerned that the White House had dropped the communications ball.

## NOTE

[1] I/M was only one of many items on crowded state legislative agendas. The Illinois 1994 legislative agenda illustrates the context in which enhanced I/M states deliberated over EPA's requirements. Other top-priority issues facing Illinois legislators in 1994 included mandatory sentencing for crimes, gun control options, whether to allow riverboat gambling, state procurement ethics policies, budget process reforms, and employer trip reduction programs (Halperin 1994).

CHAPTER 5

# THE DEAL: FLEXIBILITY GRANTED

On January 10, 1994, we heard a rumor from Carla Anderson in Senator Presley's office that the legislature was now going to move SB 629 to a vote, in retaliation for starting the sanctions process. As press reports made obvious, the sanctions announcement had all but ended our hopes for a negotiated settlement: "Negotiations ... over the smog program collapsed after last week's inexplicable decision by the EPA to move ahead on sanctions. ... If EPA's miscalculation was meant to get state officials to roll over, it was a flop. They haven't, and now they're determined to turn this into a states' rights issue" (*San Diego Union-Tribune* 1994).

The press focused on our failed use of sanctions to motivate California, when in fact that was only part of the story. Headquarters initiated sanctions primarily to motivate the other states to keep in line, not to help us reach a California-EPA compromise. Regardless of the motivation, however, the sanctions announcement threw the Region 9 negotiation process out the window.

## THE GROUND MOVES: JANUARY 1994

Monday, January 17, was a federal holiday, Martin Luther King Jr Day, and our entire team had the day off. Mother Nature, however, had not taken a vacation. At 4:31 a.m. California time, a powerful earthquake, initially reported as magnitude 6.6, hit Los Angeles. Damage was severe. Even in disaster-savvy California, the Northridge earthquake was a major disruptive event. The earthquake closed sections of four freeways, and in car-dependent Los Angeles, it would take four months to reopen the most heavily damaged roads. The collapsed interchange between Interstate 5 and State Highway 14 (Figure 5-1), called "one of the most spectacular and costliest damage sites," symbolized the earthquake's destructive force (USGS 1996; Schmitt 1998).

Early estimates put financial losses well over a billion dollars. A mere 10 days after EPA announced plans to sanction California and withhold highway funds, the president of the United States appeared on national television and promised to help California by *expediting* the delivery of hundreds of millions of dollars in federal highway funds. In one of many Smog Check ironies, the epicenter of the quake was located in the heart of Richard Katz's state assembly district.

## REVERSE TRIGGER: JANUARY–FEBRUARY 1994

The state legislature passed SB 629 on January 19. But despite passage of what EPA considered to be a "bad bill," the earthquake enabled us to get back into negotiations. On January 24, Carol Browner sent a letter to Governor Wilson notifying him that because of the earthquake, EPA planned to cancel the "recently announced accelerated deadline for imposing sanctions" (Browner 1994a). Katz responded favorably. As the *Los Angeles Times* reported, "Richard Katz (D-Sylmar), a lead negotiator and architect of the California [Smog Check] bill, said that 'given the fact that the EPA has backed off, I think the appropriate response is to take another run at getting a solution'" (1994b).

Katz's quick and positive reaction to Browner's decision to halt sanctions gave those of us in Region 9 hope we could find a compromise. What the California press did not focus on, though, was the risk such a compromise would pose to the adoption of test-only programs in other states.

### The Fifth Negotiating Session: January 26, 1994

On January 26, I met with Felicia and Browner's chief of staff, Mike Vandenburgh, in Sacramento, where we had lunch with Katz and Stevens before joining other

**Figure 5-1. Collapsed I-5/SH 14 Interchange after the January 17, 1994, Northridge Earthquake**
*Source:* Courtesy of the USGS (1996)

state officials for negotiations. Katz said everybody knew now that EPA was never going to sanction California. He said that the governor was going to sign SB 629, but a companion bill was needed because SB 629 did not have an urgency clause (a clause that would allow legislation to take effect immediately, rather than the following January). Legislators were prepared to talk with EPA now, because Browner had pulled away from the sanctions. But the window was very narrow.

Amazingly, the negotiating session was essentially the principals' meeting that had failed to take place on January 11. Present were CalEPA secretary Jim Strock; legislators Dan Boatwright, Richard Katz, Quentin Kopp, Dan McCorquodale, Newton Russell, and Byron Sher; plus staff. Initially, Kopp and Russell appeared aloof, probably because their bill had already passed. But Felicia, as Mike Vandenburgh later told Carol Browner, "filibustered her way through for the first hour," talking nonstop until Kopp and Russell and the others began to grasp what Option E meant and understand that EPA had moved significantly compared with our original position. Katz was extremely helpful. Three and a half hours later, we agreed the meeting had been very productive and we should continue negotiations.

By midday Thursday, January 27, the state's technical staff had given EPA a proposal to run a pilot study that would examine the technical issues still up for debate. At about 6:00 p.m., Governor Wilson let the senate know that he had signed SB 629. In his written communication, the governor noted that "recent meetings with the U.S. EPA representatives indicate the potential for working out additional modifications to SB 629. I am encouraged by these meetings and am hopeful that all the parties involved will be able to continue to work together towards a satisfactory agreement" (Wilson 1994).

## The Sixth Negotiating Session: January 28, 1994

As positive as things had appeared on Thursday evening, the floor fell out the next morning. Dave Howekamp and I spoke with Ann Arbor staff on Friday to sort through their concerns over the pilot studies, and they had many. We called CARB and BAR staff later in the morning, and Gene Tierney relayed Ann Arbor's comments. The call turned confrontational; state staff complained that we were not being constructive. Following the call, Dave and I headed to Sacramento for another round of negotiations. After we circled around the core issues yet again, both sides had an epiphany: we realized we had to bring the technical staff face to face. We needed to either resolve the technical issues once and for all or acknowledge we could not find common ground. We agreed to a technical negotiation to take place four days later, in Washington, D.C.

## Bridging the Gap: Tuesday, February 1, 1994, Technical Negotiation

Early on Tuesday, we arrived at EPA's national headquarters in Washington, D.C. "Since no one else volunteered, I said I'd facilitate the meeting," I began. We collectively agreed that our goal was to learn whether any pilot program would be acceptable to both EPA and the state, and if so, whether such a pilot effort was doable in the time available.

By the meeting's end, surprisingly, we had found common ground at last. We agreed to complete two groundbreaking pilot studies. The first would compare competing dynamometer tests to learn whether the less expensive ASM test could replace EPA's IM240 test. The second would study different methods to identify the vehicles most responsible for pollution; participants agreed to run an RSD pilot program in Sacramento, plus create and test a computer model to predict gross polluters. California would take the lead, but EPA would be an active partner—tracking the work and helping resolve problems.

We also agreed that the state, with support from EPA, would finish the pilot work by December 31, 1994, 11 months away. The deadline was driven by EPA regulations requiring program phase-in beginning on January 1, 1995. With the pilot study results in hand, the state would know what program elements to implement during 1995.

The group then defined three remaining steps to complete a deal. First, we needed to have an MOA spelling out the commitments. Second, attached to the MOA, we needed a protocol describing the pilot studies. Finally, we needed the state legislature and the governor to authorize the program.

"I heard," said Katz's aide, John Stevens, "that we've gone through all the technical issues and found that we have conceptual agreement on how to do this. Granted, there are details to be worked out, but we can report back to our bosses there are no deal breakers." Everyone concurred that we had reached agreement on how to implement the pilot studies. We had yet to resolve some of the core policy issues, including the applicability of the 50 percent discount and the state commitment to a reverse trigger. But we had made important progress.

## The Seventh Negotiating Session: February 2, 1994

On February 2, Dave, Felicia, and I went to Sacramento for our next session. Katz, again being helpful, flagged political obstacles and ways to maneuver around them. He emphasized how little time was available: "We have a bipartisan delegation visiting Washington February 22 to 24. I really want this all done before the Washington trip." We closed feeling as though we were on the fast track toward a resolution.

## The Eighth Negotiating Session: February 4, 1994, Staff Meeting

At our next staff-level negotiating session two days later, however, we seemed to take several steps backward. Senator Kopp's staff questioned the meaning of a reverse trigger (Katz's plan to commit to meet EPA requirements, then back away as state alternatives proved effective). CalEPA raised concerns about the 50 percent discount. At the end of the day, it seemed we had less than when we started.

## NEGOTIATION FATIGUE: FEBRUARY 1994

### The Ninth Negotiating Session: February 7, 1994

On February 7, Dave and I returned to Sacramento. By the end of the meeting, CalEPA agreed to give us a specific proposal to study the 50 percent discount. We also spent time debating the substance of the reverse trigger, to which CalEPA had not yet committed. As Dave and I returned to San Francisco, we were cautiously optimistic that if we could work through the 50 percent discount and get CalEPA on board with the reverse trigger, we might have a deal.

News reports that day helped fuel our interest in quickly reaching agreement; nationally, EPA was losing support among elected officials, and the lingering I/M controversy was certainly not improving the agency's reputation: "Antiregulation forces in Congress have won their first big skirmish of the year, derailing a bill to elevate the Environmental Protection Agency to a Cabinet department. The House vote on Wednesday ... caught Democratic leaders off guard" (*New York Times* 1994). The news reports provided an important reminder to those of us at EPA that despite the high environmental expectations that accompanied the election of Clinton and Gore, other national-level policymakers were critical of costly environmental regulations, particularly those that involved unfunded federal mandates, and were prepared to place limits on EPA's authority.

### The Emerging Deal

On February 8, we held a national EPA-only conference call to brief the other regions about the impending deal. We explained that California was committing to do the following:

- Implement Option E, which meant that vehicles six years old and older—60 percent of the fleet—would go to test-only facilities; newer vehicles could go to test-and-repair stations. Because California tested cars every two years, 30 percent of the fleet would go to test-only each year.
- Use the IM240 test, unless alternatives proved effective. This was a key element of the reverse trigger.

- Complete pilot studies, by year-end and with EPA support, to evaluate RSD and gross-polluter identification software and to compare ASM with IM240.
- Meet EPA's implementation deadlines: have the ability to test 15 percent of the fleet in test-only facilities by January 1, 1995, and fully implement a program by January 1, 1996 (the final design contingent on pilot study results).

With a deal imminent, we anticipated that Mary Nichols and Carol Browner's chief of staff, Mike Vandenburgh, would fly to California the next day. To tie up loose ends, Dave and I had another negotiating session in Sacramento.

### The Tenth Negotiating Session: Afternoon of February 8, 1994

At the end of negotiations, John Stevens recapped outstanding concerns. They included three core issues: the reverse trigger, the 50 percent discount, and EPA flexibility on deadlines (CalEPA wanted some assurance that EPA would not start sanctions if the pilot studies took longer than expected). Some lesser issues were also on the table, such as whether car washes and gasoline sales were allowable at test-only stations.

### The Eleventh Negotiating Session: The Principals' Meeting, February 9, 1994

On February 9, Mary and Mike arrived from Washington, and we filed into a legislative hearing room for the long-awaited principals' meeting. Present were Assemblymen Sher, Katz, and Andal; Senators Kopp, Russell, and Presley; CalEPA secretary Jim Strock and deputy secretary Mike Kahoe; and other staff. John Stevens circulated a list of the outstanding issues that separated us from a deal.

First on Stevens's list was the 50 percent discount. We explained that EPA did not know how to study this issue in the time available for the pilot studies. After making no progress here, we moved on to the next issue, discussion of the reverse trigger. To start the discussion, Katz asked Strock whether he supported the reverse trigger: "Jim, what about you?" Strock did not give an answer. Instead, he talked about the importance of the other issues. At this point, Katz was at his best: "Well, let me see if I hear what you're telling me. What you're saying is that if you get agreement on these other key issues, then the trigger falls into place. That's not a problem for you, then?" At this, Strock nodded yes. Finally we were getting somewhere. Then Katz asked Kopp what he thought. "My position hasn't changed," said Kopp, who opposed the reverse trigger.

At this point, Katz changed tacks. "Let's start at the bottom of the list instead and work our way up; those issues are easier." We went through the non-deal-breaker issues, and Mary quickly agreed to the state's requests. We then moved to the deadline issue, the state's request that EPA not restart the sanctions process if the pilot studies finished late. Mary committed to acknowledge good-faith efforts. Strock pressed for a more specific commitment, and then Katz became irritable. "Guys, guys, let's calm down," said Senator Russell. "We've made so much progress, and Richard, that's largely due to you; you've been the one who's brought us this far." "I apologize for doing that," Katz joked, and we all enjoyed a rare Smog Check laugh.

"Well, that leaves the 50 percent discount," Katz said. Looking at Strock, Katz asked whether, if we could settle the 50 percent discount, Strock would go along with the reverse trigger. Strock concurred. We then agreed we had done all we could in this meeting, resolving essentially all but one issue: the 50 percent discount.

As we were leaving the room, Stevens asked me to join a conversation with staff from BAR, the agency that ran Smog Check. "They say they spoke to Ann Arbor, and the staff there told them the January 1, 1995, deadline wasn't firm," said Stevens. "No," I said, "you have to have the capacity to test 15 percent of the vehicle fleet on January 1, 1995." BAR staff disagreed, and I promised to contact Ann Arbor to get an answer the next day. Little did I realize how controversial this issue would become.

## Katz Loses Patience

February 10 was frenetic. We spent the day rewriting the MOA with help from Ann Arbor and EPA's attorneys. We also tracked down an answer to the January 1, 1995, deadline question.

Around midmorning, I received confirmation from Ann Arbor that, indeed, the January 1, 1995, deadline was flexible. "How am I supposed to communicate that to California?" I complained to Dick's deputy. "We've said all along they need to have 15 percent testing capacity on January 1." "I know," he said, "I just heard this myself." It turned out that the previous October, Ann Arbor had issued a little-known memorandum giving states more flexibility on the phase-in date. I called Stevens to let him know. His initial reaction was "Great, that's a deal closer."

Later in the afternoon, though, all hell started to break loose. Stevens called to say the deadline issue derailed everything. "They are going to want to wait now and collect more data on the 50 percent discount," he told me. I broke free from the phone call, briefed Felicia and Dave, and then caught up again with Stevens. "My boss is *really* upset at you guys," he said. "He feels that all along you've said you couldn't change the regulations, and now this looks like you've changed the regulations."

By now it was 5:00 p.m. West Coast time, and we had a full-scale crisis brewing. I raced back to Felicia's office. She and Dave were on a conference call with the whole group back east; they asked me to summarize what was going on. I told them about Katz's concern: "They're now looking at this as a way to gain more time and test out the full effectiveness of SB 629. If it works, they won't ever have to build extra test-only facilities."

Afterward, together with Mike Vandenbergh, Felicia called an exasperated Katz. "I wanted to try to explain where we are on this January 1, 1995, deadline issue," she began. Katz said he had just spoken with Mary, and when he told Mary that California deserved to get a flexible deadline if other states were getting one, she agreed with him. "You have no credibility. How am I supposed to believe anything you say?" complained Katz. Felicia said she understood he was mad and added that she was mad too. She and Mike said they would talk some more with others at EPA and then call him back.

After the call, we spoke with Mike Vandenbergh again and tried to pull the pieces together. It was very late on the East Coast, and Mike had just been to California and back in the past 24 hours. Everyone was exhausted, and the appointees were justifiably upset that communications among the career staff had obviously broken down, as this was the first time we knew about such an important change to the I/M implementation deadlines. I faxed to Mike EPA's October 1992 final I/M regulations and the October 1993 memo providing implementation flexibility during 1995. We linked Mary and Dick on the line and sorted through options. As the night wore on, Mary, Mike, and Felicia, all lawyers, read the regulations and felt they could justify to Katz the requirement for 15 percent test-only capacity on January 1, 1995. Then it became a matter of distancing ourselves from the obscure October 1993 memo. We agreed to call it a night, leaving Mike and Mary to reach Katz in the morning.

By the next morning, miraculously, Felicia heard from Mike that he had reached Katz and had somehow managed to smooth things over. The tension of closing the deal was straining everyone. After sleeping on the problem overnight, Katz and Mike agreed it was time to bring Smog Check to closure. In the days that followed, the press began to sense that at long last, a deal was imminent: "After months of acrimony, state and federal officials are close to agreement on a complex new Smog Check program. ... To satisfy EPA, a large number of vehicles—including, hopefully, the worst polluters—would be sent to test-only centers. To satisfy the state, the garages would still test many vehicles" (*San Jose Mercury News* 1994).

In retrospect, the paper's observation about the complexity of the emerging deal was farsighted. At the time, those of us in Region 9 focused more on the need to get agreement on environmentally sound improvements; implementation complexity was a lesser concern.

During the week of February 14, we worked, with little success, to resolve the final issues. By midweek, the lack of progress was wearing Katz's remaining patience thin. "Let's get a deal," he told me, adding he was in this "for about another 48 hours and that's it." A key recurring problem was that EPA saw no practical way to evaluate the 50 percent discount in the short time available for the planned pilot studies. Yet the Californians, especially CalEPA, wanted to examine the discount's technical basis.

"Look," Katz said to me during one of our discussions, "what I'm telling folks over here in the capital is that if you really believe in everything that SB 629 said it would do, then you shouldn't have any trouble committing to prove it or do EPA's program." Likewise, he added, "if EPA believes the 50 percent discount is valid, then your technical staff should have nothing to fear from a test of the discount's validity." Katz's argument was simple and compelling. "The more I hear that your Ann Arbor staff doesn't know how to study the 50 percent discount issue, the more suspect I am that they don't have the data to back it up," he said.

## SOLVING THE 50 PERCENT DISCOUNT PROBLEM: FEBRUARY–MARCH 1994

Later, while visiting Washington, D.C., Katz and other California legislators met with Carol Browner, Mike Vandenbergh, and other EPA staff. Mike said that when the group broke up, they agreed to participate in one final technical meeting, in California, with me again acting as facilitator. This time, the meeting would focus solely on the 50 percent discount.

### The Twelfth Negotiating Session: February 25, 1994, the 50 Percent Discount

This negotiation, to try once and for all to resolve the 50 percent discount issue, involved Mary's aide, Cecilia Estolano, and Browner's aide, Nancy Sutley; our usual Sacramento negotiating partners; and key technical experts from EPA and the state. After about an hour and a half of various technical parries, one of CARB's lead managers succinctly stated what we all knew: we could not assess the 50 percent discount in the short time available to conduct the pilot study. Shortly after CARB's pronouncement, and sensing a meltdown in the negotiations, Nancy called for a break.

Nancy and Cecilia then had a sidebar discussion with John Stevens and CalEPA's Mike Kahoe. They struck upon a creative solution that moved us through the crisis by deflecting attention away from the 50 percent discount. Their insight was that the use of RSD and the state's other innovations would generate real-world emission reductions, and these would offset the importance of the

50 percent discount. For example, if RSD helped find high emitters and reduced pollution, these impacts would supplement the regular Smog Check system. Therefore, even though the 50 percent discount would remain intact, any supplemental emission reductions would help offset the discount's impact. The more supplemental reductions that were created, the more cars the state could allow to use service stations instead of test-only facilities. A key assumption here was that RSD and the other I/M pilot study elements would prove successful. The shift in focus was an artful political construct; it bypassed the technical debate and went straight to the political concerns about how many vehicles would be sent to test-only stations.

Working with all the negotiation participants, we crafted a one-page memo that said the participants agreed that the results of the pilot study could affect *the impact of* the 50 percent discount. Cecilia Estolano, Mike Kahoe, John Stevens, and Nancy Sutley had finally placed us on a path that would lead us out of the debate maze.

After the negotiations ended, the EPA delegation called Felicia. "We all agreed that we can't do a real study on the 50 percent issue," Cecilia told Felicia. She explained that California agreed that the pilot studies, if successful, would generate extra credit that could be used to lower the number of test-only cars.

Then Cecilia brought up a new—and key—issue that would decide whether we had a deal: specificity. The specificity issue had not arisen during the negotiation meeting, so it was a surprise to most of us; we were hearing of it for the first time as Cecilia described it to Felicia. Cecilia explained that when she and Nancy had their sidebar discussion, Kahoe and Stevens asked how much specificity the California-EPA agreement would contain when describing the state's default commitment to Option E, the commitment to send 60 percent of cars to test-only unless the pilot studies proved successful. The state's commitment to Option E was the heart of the reverse trigger. Nancy said that Kahoe and Stevens had suggested MOA text that could say something like "all six-year-old and older vehicles, or their equivalent, would go to test-only stations." The "or their equivalent" language was a key wrinkle; it left room for the state to substitute the Option E commitment with something they considered equally effective.

## THE DEAL: MARCH 1994

### Negotiating Specificity

March 3, 1994, opened with big news from Cecilia. She had spoken with Stevens, and the specificity issue was the only remaining deal breaker. The last negotiation had moved us past the 50 percent discount once and for all. Later in the day, our technical staff finalized the pilot study protocol with California. However, there was still no agreement on the specificity text to be included in the MOA. What exactly was California committing to? What would happen if its pilot program proved a success? Was California committing to Option E or just to meet the EPA performance standard?

By Monday, March 7, no resolution had yet been reached. Mary seemed to think the state needed to commit unequivocally to meet Option E. In the afternoon, one of my staff heard that a committee vote on legislation was scheduled for Wednesday, regardless of the state of the negotiations.

Late in the day, Felicia, Dave, and I called Mike Kahoe and John Stevens to discuss the MOA once again. "OK, let's start in and walk through this," said Stevens. We got down to the heart of the deal: California's commitment. After some verbal jockeying, Stevens lost patience. "We've got a conference committee on Wednesday, and we'll see you there," he said, nearly ending the discussion. Felicia then tried to change the tone, telling Kahoe and Stevens that EPA had said all along we needed a specific commitment. "The premise from the start of these discussions, John, was that you were going to commit to something up front that EPA could approve, and then run a pilot program to see whether you could prove the usefulness of RSD and everything else, and then get to change your program to reflect that," she said. The call was breaking down fast.

While Dave and Felicia maneuvered, I looked at the draft MOA and wrote some new text. During a pause, I ventured forward. "Look," I began, "I'm not sure if our headquarters folks would buy this, but how about I read you some alternative text?" "OK," said John, sounding tired. "What I have in mind is some kind of text that says that the state commits to a program, and here's the key part: that it demonstrates to EPA's satisfaction that the program meets the emission reduction performance standard. What do you think?" I tried to deflect some of the attention away from the specific commitment but still ensure that whatever the state did, it would have to have EPA's approval. Kahoe and Stevens mulled it over, with Stevens speaking first: "I want to look at that some more, but it sounds like something that ought to be OK. Mike?" "Yeah, I think it sounds OK, but I want to think about it some more," said Kahoe. Felicia scribbled me a note and threw it across the table: "Way to change the momentum!" it said, and she threw up her hands and leaned back.

Tuesday began with Felicia explaining the specificity issue to Mike Vandenbergh. Mike said he would brief Carol Browner to let her know that EPA's decision on this issue could potentially resolve or collapse the negotiations, depending on how we handled California's request.

By midmorning, when we were supposed to phone Kahoe and Stevens, it was clear that without a decision on the specificity question, we were not yet ready to negotiate the final MOA. We were down to hours left to negotiate, with the conference committee scheduled for the next day. I left messages for both Kahoe and Stevens to tell them we were running late.

At 3:30 p.m. (6:30 on the East Coast), we had yet another call with headquarters. All the key EPA players participated. Mary passed on some decisive political news. She said that earlier, she had heard that Carla Anderson of Senator Presley's staff agreed with Katz. "I was so surprised to hear that," said Mary, "that I immediately called Carla to talk with her." She told us that Anderson spelled out more effectively than anyone else why we should accept the language Katz was offering: First, if, down the road, California didn't have the data to prove the effectiveness of its alternative program, it would have no choice but to send the cars to test-only. Second, after the pilot program, EPA would be in a better position to require California to rely more heavily on test-only, because we would then also have the data. So far, we had been fighting the state based only on our opinion that its program wouldn't work well. Third, the complexity and expense of California's program would make other states decide they did not want the same plan. Finally, Anderson argued, we would be better off with *some* program in California, rather than the *nothing* we might wind up with if we continued the negotiations. Mary said her feeling now was that we ought to move the fight's venue out of the legislature.

We were so close. We had threatened and then pulled back from sanctions; then threatened and pulled back again. We had intervened to kill what we viewed as bad legislation, and then we had seen that same legislation come back to life and become law. We had negotiated a detailed protocol to complete scientifically based pilot studies that at first seemed impossible to complete on time. We had crafted an unprecedented MOA. And now we were down to this one major obstacle: would the state agree to a performance objective or the specific program we wanted?

We gave in. Mary made the decision, which no one opposed. We would live with the state's commitment to a program *equivalent* to what we wanted, rather than to the exact program we wanted. As detailed in the final MOA, however, the state at least would have to demonstrate program effectiveness to EPA's satisfaction.[1]

At close to 6:00 p.m., we finished up our comments on the draft agreements with the state, and I shared them with Kahoe and Stevens. In the meantime, Felicia explained our decision to them. We were now nearing 6:30 p.m., and Felicia, Dave, and I called Mike Kahoe and John Stevens one last time, to finish the MOA. By around 7:00 p.m., we were headed down the home stretch. Dave pressed the mute button and joked: "Amazing how easy this is now that we caved on the commitment language!". About 20 minutes later, it was all finished. We finally had a deal.

### Wednesday, March 9, 1994: The Conference Committee Votes

On Wednesday, March 9, the legislative conference committee voted on the I/M issue. We sent one of our staff to observe the hearing. "It was a love-in to praise the flexibility of all sides," he reported back. The legislation passed 6 to 0.

Thursday morning, however, we had more problems. The morning press accounts were all wrong. The *San Francisco Chronicle* reported that California would send just 15 percent of its cars to test-only (1994b). The paper missed the fact that the 15 percent requirement applied only during the 1995 transition year. By the time I arrived at work, Felicia had already left a voice mail venting frustration that the *Los Angeles Times* had gotten the story wrong as well. Although the press accounts did not directly affect the outcome of the California deal, they did influence other states' perception of what EPA had allowed.

John Stevens called on Friday to let us know the full legislature would vote late the following week. The moment of truth had finally arrived.

## ROLL CALL: MARCH 1994

With the full legislature about to vote on the deal, word of the California-EPA agreement was widespread and having an impact on other states. For example, the *Washington Post* carried a story suggesting that Virginia could have a deal similar to California's: "Northern Virginia drivers may be able to continue taking their cars to local garages for smog tests. ... [L]ast week, the federal Environmental Protection Agency and California officials settled a long-running dispute over emissions testing with a compromise that will allow 85 percent of cars tested each year to continue going to local garages" (1994a).

The paper had incorrectly reported the deal as allowing 85 percent of cars to be sent to garages. The deal was so complex, however, that it was easy for reporters to miss the nuances. Option E meant that vehicles six years old and older—60 percent of the California fleet—would go to test-only. Because California tested cars once every two years, only 30 percent of the fleet would be test-only in any year. However, our phase-in schedule required just 15 percent of the fleet to be sent to test-only during 1995. California's final test-only commitment would depend on the pilot study results. That was a lot of material to digest, and reporters, including many who had followed Smog Check for months, were not absorbing it all.

Thursday morning, March 17, was the day I thought might never arrive: both houses of the California legislature were set to vote on an EPA-approved I/M program. Carla Anderson patched us into a live connection to the floor of the legislature, and we listened to the votes. The I/M deal was packaged in three bills: SB 521 (Presley), which included details of the I/M program commitment; AB 2018 (Katz), which covered the pilot studies; and SB 198 (Kopp), which created an old-car buyback program to help low-income motorists. What follows are excerpts from the floor debate in both state houses. The senate debate started first, with SB 521.

### Senate

*Debate and Vote on SB 521*

**President Pro Tem**: [gavels for attention] "This is a measure that has had considerable interest for your constituents, the smog control situation . . . "

**Presley**: [describes test-only] "SB 521 basically does the centralized part of it, as I say, representing 30 percent of the vehicles, and I ask an aye vote."

**Kopp**: " . . . I invite the attention of the members to the fact that this is a milestone. It represents the culmination of 16 months of serious debate, numerous hearings, and countless informal discussions. . . . And it represents, Mr President and Members, the most sensible approach, probably in the entire country, which has been achieved as a result of persistence in staying with principles of flexibility and the uniqueness of California . . . "

**Hayden**: "This is my understanding, and perhaps Senator Kopp can comment on it. The compromise that has been reached is a form of what was known as 'Option E' in the federal EPA's original strategy. . . . So, it is definitely a compromise. . . . I have to say, the compromise sacrifices clean air to the interests of the present Smog Check system. I would prefer to stand with the Lung Association . . . they are opposed to the compromise that has been worked out. . . . While they welcome innovation including the use of remote sensing, the Lung Association says there is no compelling reason to back away from a separated test and repair system. . . . They say that if this measure is adopted, which they don't support, that it's . . . unfinished business, and that we will have to monitor, and probably return to this issue within a year, to toughen the program. . . . [I]t's not good for our health, it's not good for our lungs, and for that reason I intend to vote no."

**Peace**: "I too would like to stand with the Lung Association, but unfortunately, should I choose to, we would all be standing in line forever, waiting for our inspections. . . . It was, and the federal government's proposal was, an absolutely unenforceable, unworkable, impossible to achieve concept, classic kind of academic approach, to something that clearly could not work with the millions of people and the millions of cars that we would have had to have dealt with. ...

I think too there's also a lesson here with respect to our relationship with the federal government in general. ... This is a small victory. An important one in the context of clean air. But even more important in its symbolism, and its effectiveness at establishing a precedent for the ability of this state to send the message to Washington that says we will no longer allow you to simply run over on us on things we cannot afford and cannot practically implement. We will stand, we will fight, and we will win."

**President Pro Tem**: [roll is called] "Ayes 32, Nos 3, the bill is passed to the Assembly. Now, continuing on the second of the trio we are dealing with, item number 8 on file, Katz 2018, who is having that matter? Senator Russell? Read the bill."

**Clerk**: "Assembly Bill 2018 by Assemblyman Katz an act relating to vehicles."

*Debate and Vote on AB 2018*

**Russell**: "This is part of the package that Senator Kopp, Senator Presley, and I'm carrying for Assemblyman Katz. ... This bill allows us to have our demonstration program, in conjunction with the federal EPA ... and determine how close, or how much we exceed the federal standards. If we fail, all we have to do as a result of this bill is to increase just enough to meet those standards. We don't have to go up to 100% [test-only]. ... I request an aye vote, and I think it's one you need not be concerned about if you vote for it. ..."

**President Pro Tem**: [roll is called] "Ayes 31, Nos 3, the bill is passed."

*Debate and Vote on SB 198*

**President Pro Tem**: "Item number 9 on file. Senate Bill 198, Transportation Committee, Senator Kopp, read the bill."

**Kopp**: "This bill comprises the so-called buyback program. ... This allows a car owner to pay in $50 to a special fund ... from people who want to be exempted from that first Smog Check [after buying a new car]. Now, out of that fund there's available $450 in a subsidy for repairs ... or, if you want to buy a new car, there is available up to $800. ... So that's SB 198, which is the third and final bill of this statutory scheme, which will set standards and will set the pace for the rest of the country. Of that I assure you, because we will succeed with the radar sensing, with the random checking of vehicles, establishing a system eventually, that makes much more logical sense than simply taking a photograph, so to speak, of a vehicle on one day out of two years. ..."

**President Pro Tem**: [votes are tallied] "Ayes 32, Nos 1 ... the bill is passed to the Assembly."

**Assembly**

*Debate and Vote on SB 521*

**Katz**: "... We jointly develop a demonstration program to show the effectiveness of remote sensing devices and other methods to target the dirtiest cars to send to the test-only stations and to prove that either of several types of tests and test equipment are just as good but less expensive, and therefore more available, than what EPA had originally demanded. ... We have put in place the strongest program in the country, the most creative program in the country, one that will clean up our air, be convenient for motorists, and keep as many small businesses as possible in operation. ... There has been a tremendous amount of work that has gone into this compromise. ..."

**Andal**: "... We also, frankly, have sent another message to Washington. And that message is the states are better able to solve the problems of their states than the federal government is. Many other states since this compromise has been achieved have pulled back from original agreements they were about to make with the U.S. EPA, and they've been enlivened to fight for their own state rights, to have their freedoms, and their problems corrected in their own states, not by Washington. ..."

**Speaker**: [rings the gong to vote; SB 521 passes 54 to 8, *just one vote* over the 53 needed for a two-thirds majority for an urgency bill]

*Debate and Vote on AB 2018*

**Katz**: "Mr Speaker, I'd reference the previous debate and ask that the roll be opened."

**Speaker**: "Close the roll and tally the vote. Ayes 55, Nos none, conference report is adopted."

*Debate and Vote on SB 198*

**Katz**: [discusses Kopp's program]

**Speaker**: [closes the debate and calls for a vote] "Ayes 56, Nos 2, conference report is approved."

All of the legislation was on its way to the governor.

## CHAMPAGNE AND LAW

Late in the day, Felicia delivered us a bottle of champagne. She said that after the votes, Browner called to congratulate her and the rest of our I/M team. Felicia told us that she and Mike Vandenbergh had joked about no longer having to wait for the next Smog Check crisis to hit. I felt an odd emptiness; it was going to be strange not to grapple with Smog Check every day.

On March 18, 1994, the three Smog Check bills were officially received by Governor Wilson. He had 12 days to sign or veto the bills, or let them pass into law without his signature. Although the negotiation marathon had ended for EPA's San Francisco regional office, the fallout was just gathering momentum for our counterparts around the country: "California's compromise with the federal government on revamping its Smog Check program has touched off a rebellious debate in state capitols across the nation, with many officials hoping to strike similar special deals" (*Los Angeles Times* 1994c). While newspapers were reporting on the I/M ramifications for other states, the press accounts did not anticipate the much wider impact the I/M debate was to have in the months ahead for the Clean Air Act.

On Wednesday, March 30, I received word from Senator Presley's office: Governor Wilson had signed all three bills. While I was on the phone with Presley's staff, Katz's staff called and left the same message. The saga was finally over.

## NOTE

[1] The text of the California-EPA agreement is documented in U.S. EPA 1994a.

CHAPTER 6

# TIMELINES: A BRIEF VISUALIZATION OF THE DEBATE PERIOD

This chapter pulls away from the debate perspective given in Chapters 3 through 5 and provides a high-level representation of the Smog Check conflict. The text helps readers envision the state and national context for the flow of the debate, before the rest of the book moves forward in time to track the arc of the U.S. I/M experience and derive policy lessons.

To visually link debate milestones to other contemporaneous events and economic conditions, timelines are used to illustrate the duration of different debate stages and transition points from one stage to another. As the timelines show, debate outcomes were linked to several pivotal situations or actions, some of which occurred in substantially different state and national contexts.

The period covered begins with the release of EPA's draft regulations in July 1992 and ends with California governor Wilson's signing of legislation authorizing the California-EPA agreements in March 1994. The first timeline (Figure 6-1) starts off by identifying important milestones in the Smog Check debate. Next, I reviewed more than 7,000 newspaper articles published before and during the debate to reconstruct important contemporaneous state (Figure 6-2) and national (Figure 6-3) events related to air quality and the environment. I then illustrate state

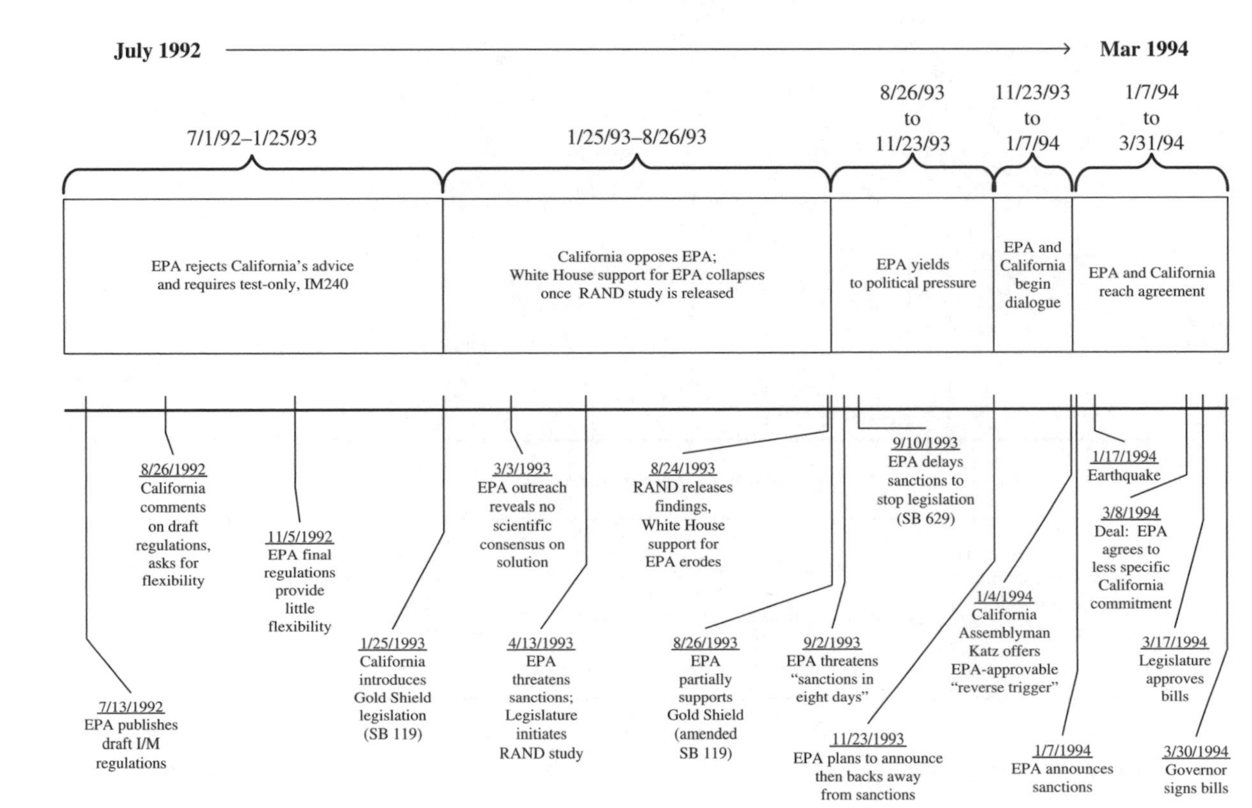


**Figure 6-1. Major Milestones During the California Enhanced I/M Negotiations**

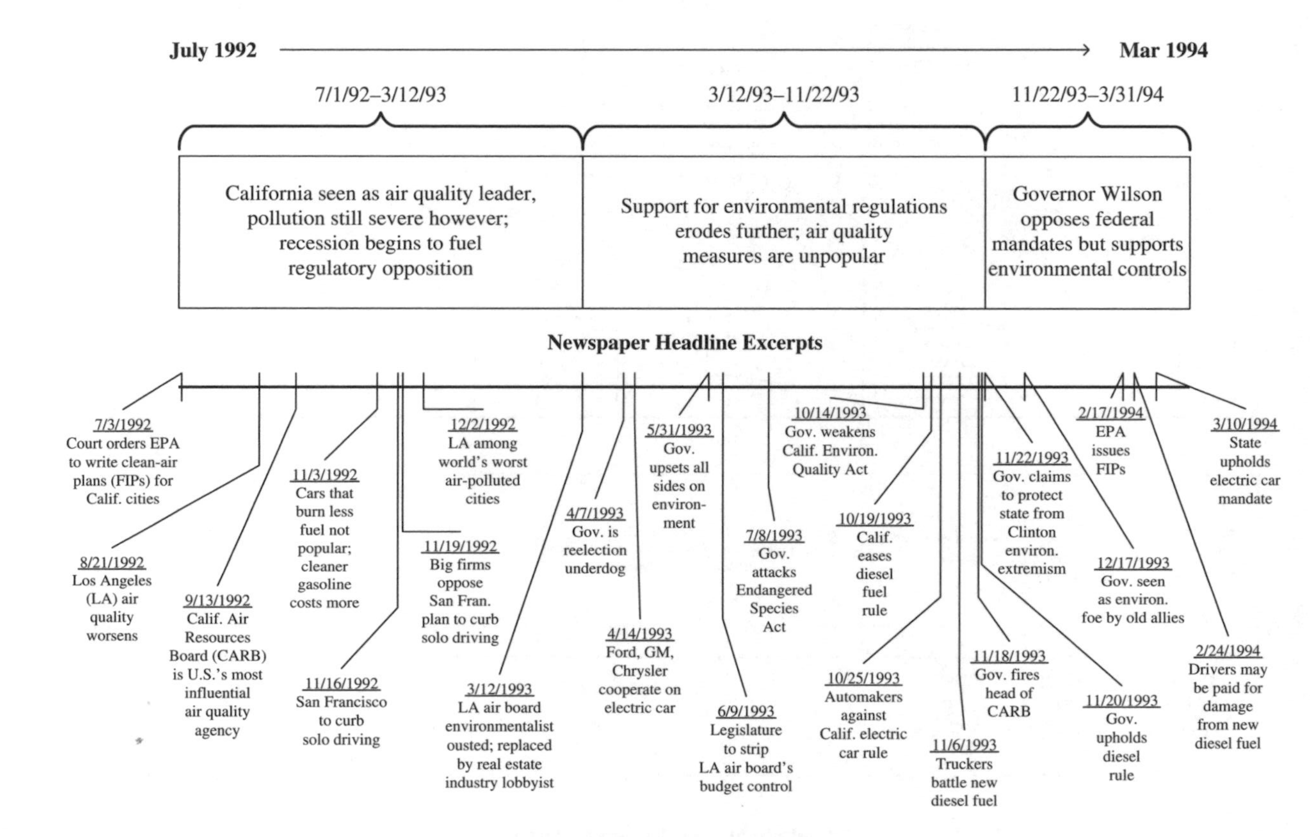


**Figure 6-2. California Air-Quality-Related Milestones Contemporaneous with the Enhanced I/M Debate**

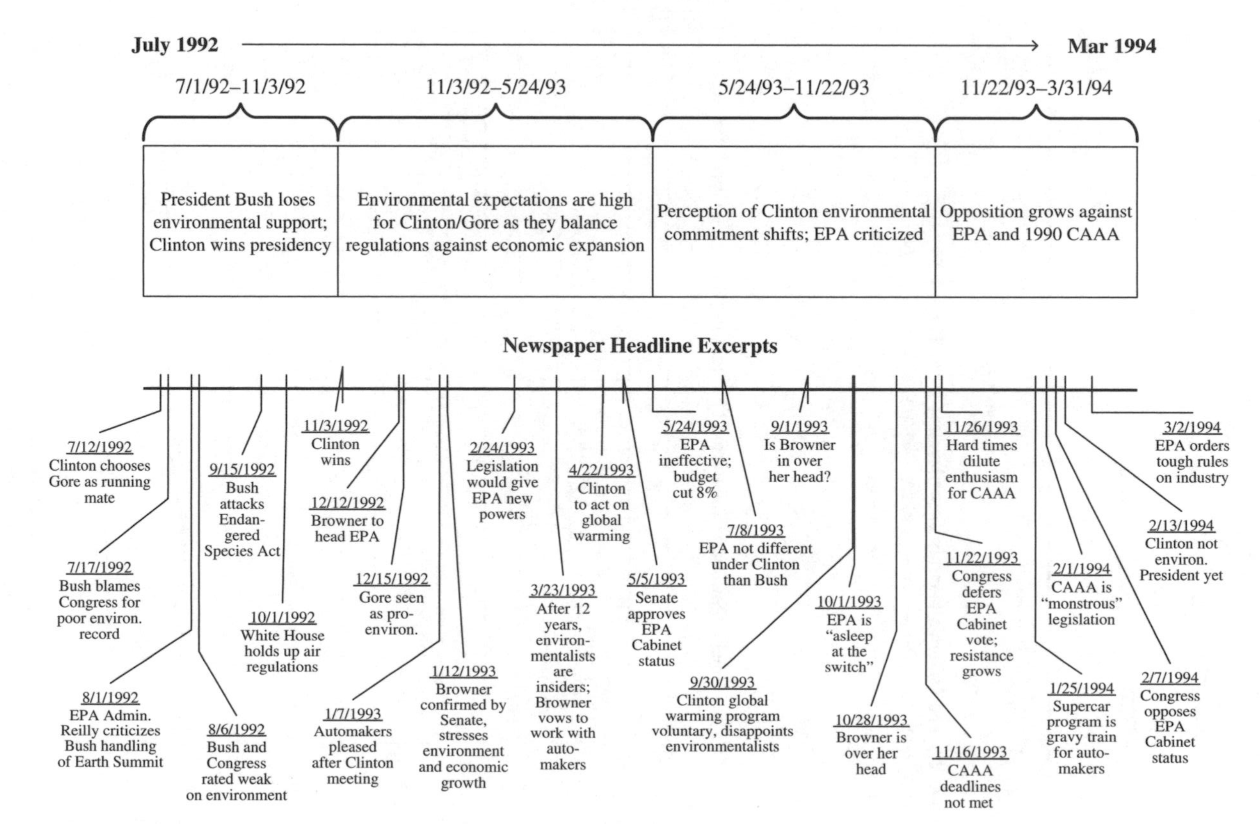


**Figure 6-3. U.S. Air-Quality-Related Milestones Contemporaneous with the Enhanced I/M Debate**

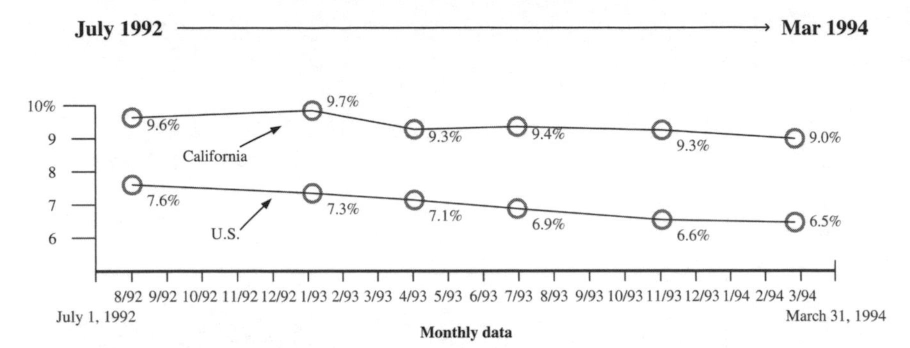


**Figure 6-4. U.S. and California Monthly Unemployment Data, 1992–1994**
*Source:* U.S. unemployment data, BLS 2003; California unemployment data, CEDD 2003

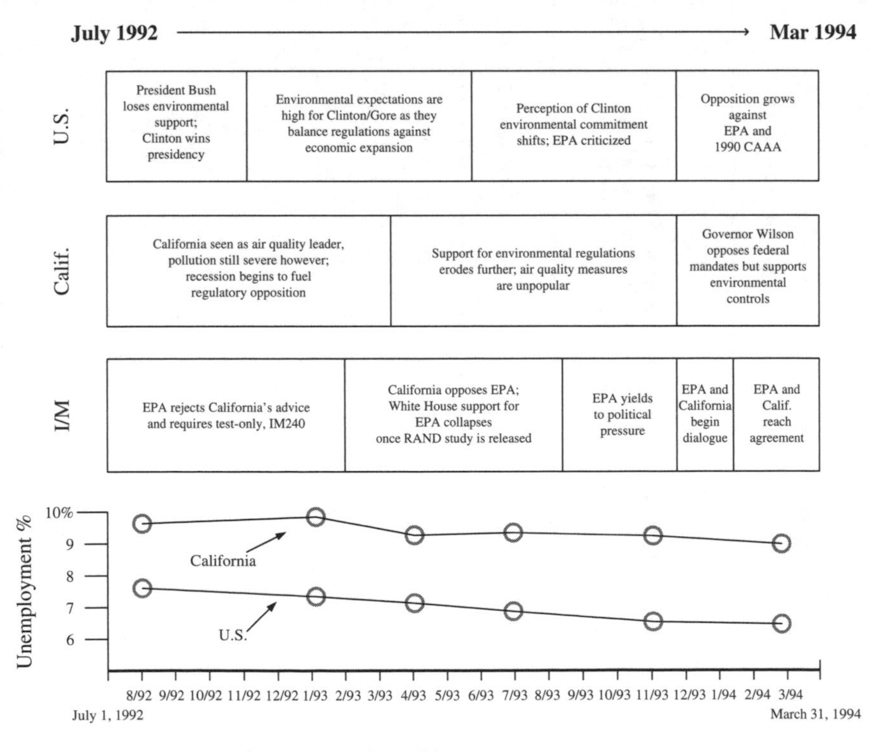


**Figure 6-5. Synthesis: Overlapping Events and Conditions**

and federal unemployment statistics (Figure 6-4), which serve as an indicator of the economic conditions during the debate. Figure 6-5 synthesizes all of the time-varying information into one graphic.

In my review of articles on then current events, I used an approach known as content analysis.[1] The analysis process was subjective, as I identified broad themes and selected excerpts of newspaper headlines that I felt exemplified those themes, and thus the figures are interpretations of historical events and represent my judgment and perceptions. Others might emphasize different themes or topics. However, by identifying and consolidating information on contemporaneous events and graphically illustrating their relationships in time, the figures should enable readers to visualize the overlap among numerous events and conditions during the debate period.

## TIMELINE SYNOPSIS

Following publication of the final I/M regulations, EPA staff had only a brief window in which to decide how to handle California's request for flexibility to implement a Gold Shield program. During this same window, national leadership transitioned from the Republicans to the Democrats, when Bill Clinton won the presidential election on November 3, 1992. The final I/M regulations were published, under court order, on November 5, 1992. California introduced I/M legislation on January 25, 1993. As the Clinton administration replaced the Bush administration, there was general acknowledgment in the press that the public expected the new administration to support strong efforts to protect the environment.

California officials, weighing what legislation to propose in January 1993, acted during a period when the state encountered difficult economic conditions. Unemployment levels were rising throughout the period when California officials were beginning discussions with EPA on how to interpret the I/M regulations, and peaked just when I/M legislation needed to be introduced in the California legislature. The national economy was also in recession, but this was less pronounced than in California and apparently was not worsening, using unemployment conditions as an indicator.

EPA took its least flexible stance on the I/M regulations at a time when most perceived that the Clinton administration would support strong environmental policies. However, during the middle part of 1993, when EPA attempted to maintain a tough message regarding I/M policy, perception of the new administration shifted as it became evident that it had yet to be as environmentally proactive as expected.

In addition, EPA held to its least flexible stance as support for strong environmental regulations eroded throughout California. The inflexibility could not have come at a worse time. Common sense dictates what has been noted in the literature: "There is agreement that governments are less likely to favor stringent environmental regulations during an economic crisis" (Wurzel 2002).

Given the conditions that prevailed midway through 1993, political pressure mounted on EPA to relax its program requirements. The August 1993 publication of the RAND report effectively ended EPA's enhanced I/M policy. RAND revealed the absence of any technical consensus in support of EPA's position—an outcome that reinforced the political opposition against the agency. Once the political and technical disagreements were clear, Clinton administration support for EPA's position evaporated. The RAND findings triggered a new debate phase, when the merger between technical and political problems forced EPA to abandon its position and yield to requests to compromise.

The final I/M negotiation phases took place against the backdrop of the upcoming California gubernatorial election, set for November 1994. By early 1994, California governor Pete Wilson, then viewed as an election underdog, and beset by Republican and Democratic critics who advocated for stronger environmental protection, softened his antienvironmental positions. The confluence of state interest in reaching a negotiated settlement, combined with increased EPA flexibility, resolved the dispute and opened the way to begin pilot projects and subsequent full program implementation.

## A NOTE ABOUT THE EARLY 1990s RECESSION

Around the start of 2008, the United States and other countries around the world began to experience a recession that was more severe than any economic downturn since the Great Depression of the early twentieth century. The dramatic scale of the worldwide economic crisis dwarfed the economic difficulties encountered in many earlier downturns. It may be difficult, therefore, years after the fact, to grasp the relative importance of the economic recession that took place during the Smog Check debate period. That recession lasted 34 months and was, at the time, California's longest economic downturn in nearly four decades (CA DOF 2008). Figure 6-6 provides broader historical context to the unemployment conditions during the Smog Check debate; it reinforces understanding about why California's elected officials were especially concerned at the time about environmental regulations that could potentially reduce jobs and increase consumer costs.

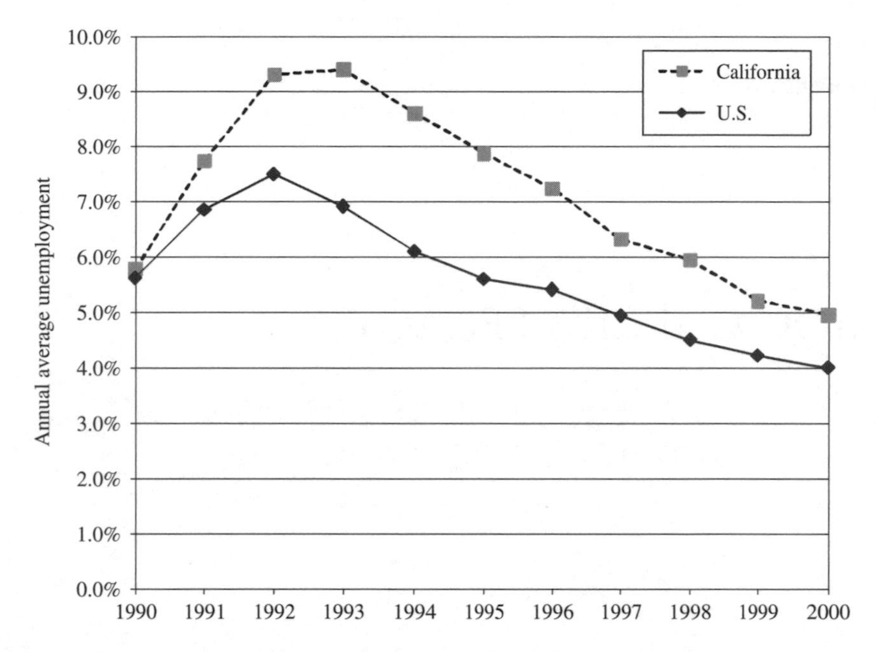


**Figure 6-6. U.S. and California Annual Average Unemployment Data, 1990–2000**
*Source:* U.S. unemployment data, BLS 2003; California unemployment data, CEDD 2003

## NOTE

[1] One definition of content analysis is "a method of analysis used in qualitative research in which text (notes) are systematically examined by identifying and grouping themes and coding, classifying and developing categories" (CiREM 2004).

CHAPTER 7

# SMOG CHECK'S LEGACY

What happened after the Smog Check conflict? If the debate and its resolution had affected only California, the outcomes might have been limited. In reality, however, the debate over Smog Check had long-lasting repercussions not just for California, but for other states and EPA.

In the short term, the California-EPA conflict brought to a head state frustration over the agency's handling of the 1990 CAAA and sparked a movement to change the Clean Air Act itself. Although EPA's senior leaders were able to save the act, their efforts required deft political and policy maneuvering. In California, the innovative pilot studies agreed to as part of the California-EPA deal wound up finding that neither EPA's model program nor California's RSD and Gold Shield–based efforts proved as effective as hoped. Across the United States, the unraveling of EPA's 1992 policy fostered state experimentation, led to disparate policy outcomes, and caused implementation delays. The resulting I/M program benefits fell far short of original expectations.

Over the long term, the consensus view among air quality professionals was that I/M programs remained a key air quality management tool, despite their many political and technical problems and the fact that they were not as effective as some had hoped in the early 1990s. Notably, nearly a decade following the Smog Check dispute and its aftermath, the view among many air quality professionals in the United States and abroad was that the core tenet of EPA's 1992 policy—the separation of testing from repair work—turned out to be correct. By the time this consensus was apparent, however, EPA had long since moved away from the 1992 policy that had created such political and technical friction.

EPA ultimately replaced its 1992 vision with a new solution: use of on-board diagnostic (OBD) computer equipment to identify and solve vehicle emission problems. Unfortunately, in 2001, EPA mandated the use of OBD-based tests before the agency had the scientific data to understand that the new tests complemented, rather than replaced, traditional exhaust measurements. The similarity between the agency's 1992 and 2001 regulatory efforts—the promotion of a preferred policy before having needed scientific data—illustrates an important and ongoing challenge regarding EPA regulatory actions.

The material that follows tracks the evolution of I/M policy from 1994 to 2007. This chapter covers both California and national outcomes and is organized to discuss the upheaval that followed the Smog Check negotiations; the evolution of I/M policy and program oversight; and the implementation of OBD-based I/M. The discussion closes with some overall observations on the U.S. I/M policy experience.

## UPHEAVAL FOLLOWING THE CALIFORNIA-EPA DEAL

California worked to quickly implement the pilot studies in the months following the March 1994 deal. It had a strong motivation to succeed: if the pilot studies failed to show that RSD, high-emitter computer profiling, and ASM testing were effective, by 1996 California would have to send up to 60 percent of its fleet to test-only inspections—a substantial increase over the state's initial commitment of 15 percent. However, while the pilot study unfolded, unforeseen events in Maine and Washington, D.C., forced EPA to broaden the flexibility it offered states.

Maine had been one of the few states to aggressively implement an EPA test-only program. The state started centralized IM240 testing in July 1994; by September 1994, however, the program collapsed. State officials halted the program following start-up problems that resulted in long lines for motorists and reports of vehicle damage during testing. Richard Carey, a Democratic state senator from Maine, was quoted in news reports about I/M implementation as saying, "It was amateur hour." Various press publications tracked the implementation difficulties, with comments such as the following from Inside

EPA's *Mobile Source Report*: "Almost as abruptly as it began, the commotion over Maine's enhanced inspection/maintenance (I/M) program has ended with the first state to start the federal program becoming the first to suspend it after just two months ... in the middle of a public-relations disaster" (1994a).

Beyond what was reported in the press, however, the Maine episode triggered alarms throughout the I/M community about the potential public backlash that awaited agencies if implementation proceeded poorly. In response, OMS briefed the EPA regions about what went wrong, concluding that EPA and Maine had done a poor job of educating the public before the I/M program started; there was insufficient lead time to phase the program in and allow the public to adjust; an ineffective effort had been made to counteract negative press when the program began; political support was weak; and Maine suffered by being the first state to implement EPA's program (U.S. EPA 1995a).

Soon after the Maine debacle, the national political landscape shifted. November 1994 marked the midterm U.S. election cycle. On November 8, 1994, Republicans claimed control of the U.S. Congress in what some journalists dubbed the "Revolution of 1994." It was the first time in more than 40 years that the political party of a sitting U.S. president (Democrat Bill Clinton) had lost control of Congress. Within a day of the election, Newt Gingrich, a Republican leader and soon-to-be Speaker of the U.S. House of Representatives, announced a plan to pass 10 pieces of legislation within the new Congress's first 100 days; one of the bills was intended to roll back government regulations.

Following the November elections, EPA's political appointees feared a congressional backlash against the Clean Air Act. The press began reporting what many in the environmental community feared would be a worst-case outcome: "Enhanced I/M and other Clean Air Act measures—already under attack—have become the target of revived resistance against federal mandates. ... The Clean Air Act, which many doubted ever would be reopened, may be up for revision with changing political tides" (Inside EPA's *Mobile Source Report* 1994b).

What some of the press neglected to mention was that governors from both parties were especially upset over I/M. The growing opposition to I/M and other air programs prompted fast action on the part of EPA. Within weeks after the Republican takeover, EPA's senior leaders formulated a preemptive strategy to save the Clean Air Act. Governors from around the country were planning to visit Washington, D.C., in early December 1994 to attend the annual National Governors Association meeting. EPA leaders decided that the administrator should meet with several governors and announce that the agency was changing its I/M regulations.

On December 8, 1994, Browner met with the governors of Colorado, Delaware, New Jersey, Ohio, Vermont, West Virginia, and Wisconsin. Mary

Nichols later briefed the EPA regions on what took place. Mary said that Browner acknowledged to the governors that portions of the Clean Air Act had been controversial and difficult for states to implement, especially the I/M provisions. Browner said EPA would reconsider the best way to implement I/M so that the program made economic and environmental sense. The response, said Mary, was "extremely enthusiastic."

Press reports following Browner's meeting with the governors underscored for EPA's senior leaders that the threat to the Clean Air Act was real, and that the agency needed to show more regulatory flexibility to defuse the anger of elected officials. As the *Washington Post* reported: "Bowing to a growing rebellion among the nation's governors, the Environmental Protection Agency is backing down from tough new auto pollution tests ... the National Governors Association said ... Browner 'is scared to death the lid is going to blow off this act [CAAA] in Congress, and she has reason to be'" (1994b). But the problem was that greater flexibility placed some states in a difficult spot.

States that had already committed to EPA's test-only program faced political and policy upheaval following the California compromise and EPA's subsequent commitment to provide greater flexibility. This was the case in Texas, which "could be one of the big losers—to the tune of $100 million," said the *Wall Street Journal.* "EPA's new flexibility appears to have left the state with no good choices. ... If the state changes the program ... then the state must compensate the two testing companies with which it contracted two years ago to ... do the centralized testing" (1994).

Consequently, governors sought more clarity on federal requirements, triggering EPA to begin defining on paper what the new flexibility meant. In a December 20, 1994, letter to Alabama governor Jim Folsom, Jr., Browner (1994b) outlined I/M program changes she said would "provide substantial additional flexibility," including the following:

- allowing states to avoid enhanced I/M if they demonstrated they did not need the emission reductions to achieve clean air;
- allowing states to create hybrid I/M programs like the one California envisioned;
- allowing states to have initial tests at test-only stations, with retesting at a service station (similar to California's original Gold Shield idea); and
- giving states additional credit if they chose to implement RSD.

Ann Arbor kept a state-by-state report card noting which states were implementing enhanced I/M, and by late 1994, the picture looked bleak. EPA's worst nightmares about the consequences of reaching a compromise in California now were coming to pass across the country. For example, Pennsylvania canceled its centralized program even though the Envirotest Corporation had already built the test facilities; Envirotest later won $145 million in damages from the state (PA DEP 1995).

## California Pilot Program Implementation

As EPA sent signals that it was willing to show further flexibility on I/M, California officials contacted the agency to assess the implications for the pilot studies and the MOA. Ultimately, state officials decided that regardless of the new federal flexibility, the pilot studies were still essential in determining how to improve Smog Check.

The main goal for the California RSD pilot was to deploy units throughout Sacramento to identify gross polluters; it was the first attempt to implement a full-scale real-world RSD program. During 1994, contractors deployed 10 RSD-equipped vans. Most previous studies measured emissions at just one or two sites; the pilot sought to measure emissions at 337 sites. It was an impressive effort to capture data from the region's 800,000-plus vehicles.

In addition to deploying RSD units, starting in November 1994 contractors began to recruit and IM240-test more than 3,000 vehicles, some of which had also been measured by RSD. One of the study goals was to compare RSD and IM240 test results to determine whether RSD accurately measured emissions. While RSD and IM240 testing took place, California also launched a study to identify whether ASM tests were as effective as IM240 tests. The ASM pilot involved testing approximately 600 vehicles and having about 200 of them repaired in a real-world setting.

California also sought to develop a computer program that would allow state officials to identify, in advance, vehicles most likely to be high emitters. Their plan was to evaluate millions of past Smog Check records and then create a high-emitter profile (HEP) software package to identify likely problem vehicles and send them to test-only inspections.

## Implementation Difficulties among States

As California proceeded with its pilot studies, legislators and governors in other states moved to roll back I/M requirements. By February 1995, for example, the new Texas governor (and future president), Republican George W. Bush, had signed into law the first legislation of his gubernatorial career: a bill delaying implementation of the state's centralized I/M program. "The state's much maligned vehicle emissions tests were halted late Tuesday for three months under a bill signed into law by Gov. George W. Bush. ... Mr. Bush said ... his main concern is that "Texas be treated like California or any other state, that we be granted maximum flexibility" (*Dallas Morning News* 1995).

The Texas action illustrated that the California-EPA deal and EPA's preemptive response to save the CAAA had set in motion a movement among states to design their own I/M programs. What remained unclear, however, was the program design states would choose to replace the EPA test-only model. In California, the pilot studies would help make that determination—but other states were not prepared to invest in such costly and time-consuming research.

Although the fate of I/M programs remained murky, Carol Browner's efforts to demonstrate flexibility appeared to be relieving some of the political pressure on the fate of the 1990 CAAA. By mid-February 1995, reports circulated among EPA management that congressional committee and subcommittee chairs did not support reopening the Clean Air Act. Congress would remain engaged and would question EPA vigorously on I/M, but it appeared that the agency had moved the act away from a political precipice.

## California Pilot Study Results

Given the tattered relationships from the legislative battle that had preceded the pilot studies, the mutual effort between California and EPA unfolded grudgingly at first. After months of working together, however, some of the divisions finally healed. As pilot work ended, California officials wrote to Carol Browner in March 1995: "As you know, the final phases of California's Inspection and Maintenance Pilot Projects are drawing to a close. ... All who have been involved ... are to be commended. ... This cooperative, hand-in-hand working relationship has been crucial in ensuring, as we went along, that all necessary elements of the studies were carried out as agreed upon in the MOA" (Strock and Kozberg 1995).

On March 30, 1995, BAR submitted the pilot results to EPA (Keller 1995; Klausmeier et al. 1995). CalEPA secretary James Strock followed with a press release praising the effort. He also used the occasion as a last chance to criticize EPA's original regulations: "The major difference between California's innovative plan and the federal government's cookie-cutter mandate program is that ours relies upon high technology to reduce costs and inconvenience for motorists, while also achieving comprehensive environmental protection" (CalEPA 1995).

What did the pilot studies find? During the process of arguing for their respective positions, both sides had oversold their ideas. The RSD effort, in just a few months, had gathered more than 1.3 million measurements of emissions (of any type, low to high) matched with California-registered vehicles. Of these measurements, two-thirds of them were matched with vehicles registered in

Sacramento County. Some vehicles were measured multiple times, however, and disappointingly, RSD observed less than half of the Sacramento area's vehicle fleet. Perhaps more troubling, the state's assessment concluded that RSD identified only about 20 percent of the vehicle emissions thought to exceed allowable levels. Not only did RSD miss cars, but it missed high emitters. A basic problem was that it was hard for RSD to observe vehicles driven less frequently. The study concluded, "Based on the results of the pilot program, obtaining valid [RSD] readings from 100% of the eligible vehicle population is not a realistic goal" (Klausmeier et al. 1995, *3–24*).[1]

The Sacramento pilot study also found that the preliminary HEP software was a promising method of identifying problem vehicles. However, even when RSD use was paired with the HEP, the tools still missed at least 70 percent of high-emitting vehicles. Clearly, much work needed to be done before the state could rely on either RSD or the HEP, or both used together, as the sole method of identifying problem vehicles.

Elements of EPA's model program also failed to perform as anticipated. After touting the infallibility of the IM240 test and predicting that an ASM test was a poor IM240 substitute, the pilot study results proved otherwise.[2] Overall, the study concluded that ASM provided equivalent emission reductions to IM240—in less than half the time and at about one-third the cost.

As a final blow to the agency's model program, California evaluated an alternative to the EPA's purge-and-pressure test for detecting fuel evaporation, which involved checking for leaking gas caps and hoses. Based on other studies, the state had already determined that EPA's tests were impractical and actually risked increasing emissions, because technicians sometimes broke hoses when performing the tests. The pilot study examined a new method of detecting gasoline vapors. Although the results were inconclusive, the state believed that some new test, combined with a gas cap check, would provide better emission reductions than EPA's flawed purge-and-pressure test.

Based solely on the pilot study results, California had earned the right to implement a program quite different from EPA's original regulations.

## Mounting Pressure to Reshape National Policy

In March 1995, the U.S. Congress, now under new Republican leadership, focused more attention on I/M. The House Commerce Committee held hearings to, as Congressman Joe Barton wrote to Mary Nichols, "receive testimony regarding the implementation and enforcement of the Clean Air Act Amendments of 1990. The focus of these hearings will be the effectiveness of the Inspection and Maintenance (I&M) Program" (Barton 1995). When Mary testified at the committee hearing, she defended the logic behind EPA's original regulations but also assured Congress that EPA was "in the process of modifying its I/M regulations to create a new, 'low enhanced' performance standard that will be substantially less stringent than the current 'high enhanced' performance standard" (Nichols 1995a). Numerous I/M stakeholders also testified at the hearings, including Texas state representative Jim Horn, who said:

> The EPA has consistently pushed the IM240 testing program as the most effective and, in many cases, the only method of reducing auto emissions. . . . One of the biggest complaints I have heard from my constituents is that the IM240 test is not convenient. According to some of my constituents, they had to wait in line for over an hour to have their car tested. . . . Any program, such as the EPA-mandated centralized IM240 tests, embarked upon without the public support, is certainly doomed to failure and we can all agree that failure in the area of improved air quality is an option none of us can live with. (Horn 1995)

As EPA reassured Congress that it would sensibly implement the 1990 CAAA, the agency tried to put out the I/M brush fires sweeping across the states. For example, as the end of the three-month program delay in Texas drew to a close, one of the most politically viable bills working its way through the Texas senate was a proposal to conduct annual testing at gas stations. The test itself would be relatively simple; it would not use a dynamometer to mimic real-world driving. The planned test effectively was identical to the old Smog Check program that EPA and California had labored to improve or replace.

Worried that this weak proposal would become law, Mary Nichols wrote to Governor Bush on April 11, 1995. Her letter was blunt: "The State has choices to make," said the EPA assistant administrator. "If the Senate plan is adopted, Texas will have to find additional emissions reductions to make up for the shortfall in pollution reduction resulting from a less effective I/M program" (Nichols 1995b). The governor's response was immediate and blunt in return. Writing to Carol Browner the next day, Bush said:

> I strongly object to your agency's efforts to meddle in matters being considered by the Texas Legislature. . . . For a federal agency to inject itself in the midst of our state legislative debate is uncalled for and totally inappropriate. As you may know, I have frequently said Texans can run Texas. The tone of your letter is an example of the heavy-handed approach of the federal bureaucracy, an approach which is strongly resented by the people of Texas. (Bush 1995)

The California deal for a hybrid program, the Republican takeover of Congress, the start-up problems experienced by Maine, the rollback of existing programs in states like Pennsylvania and Texas, and the threatened unraveling of the Clean Air Act all were taking their toll on EPA's I/M policy. The dam had broken, and what followed was a steady reshaping of I/M requirements beginning in September 1995 and continuing for several years. The IM240 test-only template was gone. In its place emerged—slowly at first, but inevitably—freedom for states to design whatever I/M program each chose.

EPA, in the first of several such actions, modified its enhanced I/M regulations on September 18, 1995, acknowledging for the first time in regulatory form that states had more flexibility to design programs. The September 1995 rulemaking established a low-enhanced performance standard that could be met with what EPA termed a "comprehensive decentralized, test-and-repair program." The low-enhanced program was based on idle emissions tests—the old-style testing California had used since the 1980s—rather than dynamometer tests. EPA referred to the original, 1992 performance standard as high-enhanced I/M (U.S. EPA 1995d).

Unsatisfied with the new flexibility offered by EPA, the U.S. Congress took further action. On November 28, President Clinton signed into law the National Highway System Designation Act of 1995. Congress used the bill as an opportunity to enact several important I/M policy changes, including prohibiting EPA both from requiring states to adopt or implement a test-only IM240 program and from automatically discounting by 50 percent the emission reduction credits assigned to test-and-repair programs.

On July 25, 1996, in its second major initiative to transform I/M, EPA created a new policy option specific to the northeastern United States. Because of topography and meteorology, air pollution spreads across many states in the Northeast, and the 1990 CAAA designated the area as the ozone transport region (OTR). EPA created an I/M program option called the OTR low-enhanced performance standard. These requirements were even less stringent than the low-enhanced program created in September 1995 (U.S. EPA 1996a).

## The New Smog Check Program Begins in California

Thanks to EPA's new regulatory flexibility, California no longer had to race to launch a restructured Smog Check by January 1, 1996. Instead, California began implementing pieces of its new system in mid-1996; the first step involved sending likely gross polluters to test-only stations.

Unfortunately, the new program immediately ran into difficulties: the state had underestimated the demands that would be placed on its test-only system. Motorists with vehicles identified as likely gross polluters waited days or longer to get telephone instructions on how to complete their Smog Check. The few test-only facilities available were overwhelmed, and BAR, under siege, suspended parts of the program (Carlisle 2004). Press attention focused on Smog Check's problems and helped fuel public outrage. The *San Francisco Chronicle*, for example, reported that "California's tough new smog check program has been strict . . . so strict that state officials and angry drivers are now scrambling to deal with a mammoth backlog. . . . At some of the . . . testing centers . . . car and truck owners have had to wait a month for an appointment" (1996).

Following the start-up problems, in 1996 the legislature passed AB 2515, which authorized a pilot program to certify specially qualified gas stations as gross polluter certification (GPC) facilities. At GPC stations, motorists could take high-emitting vehicles to be tested, repaired, and certified without having to go to a test-only facility. The program was similar in concept to the Gold Shield idea (Bowler 1996).

During 1997, the legislature further adjusted Smog Check to improve its public acceptability:

- AB 208 authorized repair expense waivers for gross polluters (previously, under the California-EPA deal, the state required gross polluters to repair emission problems regardless of cost); it also made it easier for low-income consumers to receive a repair cost waiver (Migden 1997).
- AB 1492 removed the Smog Check requirement for vehicles four years old and newer (Baugh 1997).
- SB 42 eliminated 1973 and older vehicles from the Smog Check program, based on the premise that even though individual older vehicles might be higher-emitting, in the aggregate few remained, and they were driven less frequently than newer cars (Kopp 1997).

As legislative tinkering continued, BAR implemented other elements of the new program, including modest RSD use. RSD piloting took place throughout the state in 1997, not so much to identify gross polluters as to familiarize the public with the roadside sensors. "A high-tech smog-sensing machine was taking aim at Marin motorists yesterday," reported the *Marin Independent Journal*. "It's the latest weapon on the state's war on smog, and local motorists should expect to see it parked along busy Marin streets four to five times a year" (1997).

The press incorrectly forecast that RSD would become a regular part of the Smog Check inspection process. Instead, RSD remained on the drawing board; the state rotated a small number of the devices throughout California but, following the disappointing results from the Sacramento pilot program, never implemented the comprehensive system originally envisioned. RSD went on to make important research contributions, and five other states included it as a regular part of their I/M programs, but it failed to live up to the expectations placed on it by California legislators during the Smog Check conflict (Appendix C describes state RSD experiences; see also Eisinger and Wathern 2008).

Beginning in January 1998, BAR began full-scale use of its HEP software to identify potential high emitters and require them to use test-only stations. Except for RSD, full implementation of California's new program took place starting in June 1998—several years after EPA's original program deadlines.

How well did the new Smog Check perform? By 2000, two studies were under way to assess the program's performance. One was prepared jointly by the state agencies responsible for Smog Check and air pollution control. A second was undertaken by the IMRC. Both studies arrived at roughly the same conclusions: Smog Check had fallen short of its expectations, and by quite a large margin (CARB 2000b; CA IMRC 2000). As acknowledged by CARB, as of 1999, enhanced I/M had achieved only 51 percent of the HC emission reductions and only 22 percent of the $NO_x$ reductions committed to in the SIP. Acknowledging the shortfall in emission reductions, CARB committed to EPA that California would implement, from 2002 through 2006, additional Smog Check improvements (Kenny 2000).

## The 50 Percent Discount Revisited: California Data

During the California-EPA debate, EPA claimed that test-and-repair programs merited only 50 percent of the credit given to test-only programs; however, the agency lacked data to prove its case. Further eroding EPA's original premise was an independent analysis of U.S. programs operating before the enhanced I/M era (1985–1992); that assessment found no evidence to support a 50 percent discount (Lawson et al. 1995).

By 2000, however, California had collected data from concurrently operating enhanced test-only and test-and-repair programs. As the state documented in its Smog Check program evaluation, test-only inspections were more successful at reducing emissions than were inspections at test-and-repair stations. CARB and BAR found that "for failing vehicles, the emissions after the inspection and repair are 33 percent lower for those vehicles that went to a Test-Only station compared to those that went to a Test and Repair station for initial inspection." They also found that on average, all test-and-repair stations reduced per-vehicle HC emissions by approximately 16 percent, whereas test-only stations achieved about a 42 percent reduction (CARB 2000b, *VI-6, Table VI-4*).

Roughly accounting for the fact that the repair of high emitters produced larger emission reductions than the repair of other vehicles, test-only stations achieved about twice the level of reductions as test-and-repair stations. Based on the Smog

Check program in place as of 2000, it appeared that EPA's 50 percent discounting of test-and-repair inspections had been correct after all.

CARB and BAR learned something else of interest during the 2000 evaluation. By ranking service stations in terms of the emission reductions they achieved, the state found that the top 25 percent of service stations performed as well at reducing pollution as test-only stations. Conceptually, this meant that if the top performers could be identified and licensed, they should be given the opportunity to test, repair, and certify all vehicles. This was the very idea behind the Gold Shield concept promoted by Senator Robert Presley.

## THE EVOLUTION OF POLICY AND PROGRAM OVERSIGHT

Stung by the fallout from its I/M policy, EPA continued to distance itself from its 1992 mandates. Ironically, at the same time, some states began to reconsider the need for the more stringent program elements EPA had originally advocated. In Missouri, for example, state officials eventually determined that an idle test was insufficient to combat ozone; in April 2000, the state implemented centralized IM240 testing (State of Missouri 2007).

On July 24, 2000, EPA modified its I/M regulations once again; these changes allowed states to implement virtually any type of I/M test program. EPA's regulatory changes extended implementation deadlines, relaxed dynamometer test standards, watered down the definition of test-only, and deemphasized exhaust tests (U.S. EPA 2000a).

Then, on April 5, 2001, EPA issued regulations that moved completely away from the IM240 test-only program it had launched nearly a decade before (U.S. EPA 2001a). This action, taken just as many in the professional community began to acknowledge the efficacy of some of EPA's key 1992 proposals, effectively ended federally required tailpipe testing for modern vehicles. The agency's April 2001 decision required that I/M tests transition to use OBD equipment, rather than dynamometer-driven exhaust tests. OBD allowed mechanics to plug a hand-held electronic scanner into a computer port in the vehicle and simply download data describing what mechanical and emission problems, if any, existed (Figure 7-1). The on-board computers had become standard U.S. equipment on all passenger vehicles beginning with the 1996 model year. In concept, a technician could simply get an OBD read-out to diagnose emission-related problems, thus avoiding the need for an exhaust test.

With its April 2001 rule, the agency granted full emission reduction credit to OBD-based I/M programs, equivalent to what states would have achieved had they implemented an IM240-based program. The regulations gave states little incentive

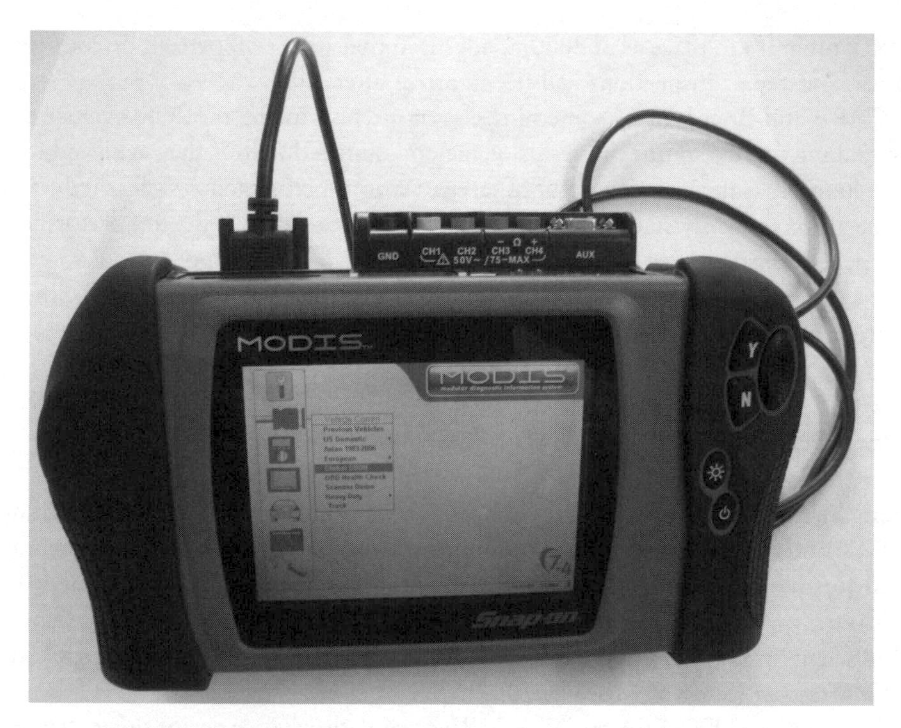


**Figure 7-1. An OBD Hand-Held Scan Tool**
*Source:* Author photo; courtesy of All Autos Inc., San Anselmo, California

to perpetuate dynamometer-based tailpipe tests for 1996 and newer vehicles. EPA required that states implement OBD-based programs beginning in 2002, although time extensions were possible.

While EPA moved to abandon exhaust tests, the U.S. National Research Council (NRC) completed an in-depth examination of I/M programs. The NRC found that because of the diversity of U.S. program types that emerged in the wake of EPA's regulatory flexibility, real-world emission reductions ranged from zero to only half of what had been predicted during the California-EPA debate period. The lowest credits were obtained by test-and-repair idle-test programs (similar to California's program in the 1980s); the highest credits were achieved by hybrid or centralized dynamometer-based test programs (similar to California's program following the 1993–1994 California-EPA deal) (NRC 2001). The NRC's concerns about lower-performing programs reinforced earlier findings that attributed poor performance to the inability of programs to control the behavior of motorists and inspection and repair professionals (Lawson 1993, 1995; Lawson et al. 1995).

Nevertheless, even though smaller than originally forecasted, emission reductions from I/M remained important to air quality management programs. As the NRC concluded in 2001, "Despite the smaller-than-forecasted benefits from I/M programs, the [I/M evaluation] committee still sees a great need for programs that repair or eliminate high-emissions vehicles" (*2–3*).

At about the same time that the NRC reaffirmed the need for I/M, an international consensus emerged supporting the central tenet of EPA's 1992 strategy: the separation of testing from repair work. In 2001 in Chongqing, China, the Asian Development Bank hosted an international workshop on "Strengthening Vehicle Inspection and Maintenance" programs and found "centralized I/M systems (sometimes called 'test only' systems), where the inspection and maintenance functions are separated, have consistently been proven much more effective than decentralized systems, where inspection and repair are combined." They also noted: "It is very difficult to supervise and audit the 'test and repair' systems and to prevent corruption and poor quality control. Policymakers must resist the adoption of programs that combine testing with repair" (Asian Development Bank 2003, *x*).

Similarly, during a 2001 gathering in Bellagio, Italy, of "representatives of nations at the forefront of motor vehicle production, consumption, and regulation," participants identified six regulatory principles to control emissions over a vehicle's lifetime, including the use of inspection and maintenance programs. They stated that "programs should separate inspection from repair, and post-inspection diagnostics should precede repair" (Energy Foundation 2001, *20*).[3]

As years of I/M program experience yielded an improved understanding of what worked, some programs retooled their efforts. Even in Texas, where legislators and then governor George W. Bush repealed a centralized dynamometer program following the California-EPA deal, officials ultimately conceded the need for a more sophisticated approach. In October 2001, after it became obvious that Houston's ozone problem was especially severe, the state legislature modified the Texas program to include dynamometer-based tests (TCEQ 2001).

## California Implements Gold Shield and Increases Test-Only

In 2002, the California legislature passed AB 2637, and Democratic governor Gray Davis signed it into law. Because of the state's ongoing pollution problems, the bill included the San Francisco Bay Area in the enhanced Smog Check program and made two other important statewide changes. First, AB 2637 exempted new vehicles for up to six years from having to be checked; the prior practice had allowed new cars to skip the test their first four years. Second, the bill allowed vehicles that had failed a test-only test to seek repair and certification at test-and-repair facilities known as Gold Shield stations. Ten years after State Senator Robert Presley had urged EPA to allow Gold Shield, California's program had come full circle.

Also in 2002, California implemented another key change: the state increased the percentage of vehicles required to have test-only inspections. After studies documented that Smog Check fell short of its emission reduction goals, California increased to 36 percent the fraction of vehicles directed to test-only stations. Later, after making further program adjustments, the state wound up sending over 60 percent of vehicles subject to Smog Check to test-only stations. As a result, by 2005, California wound up meeting the 60 percent test-only goal defined by EPA more than a decade earlier as Option E (Weisser 2006).[4]

Over time, and despite implementation challenges, studies confirmed that California's enhanced Smog Check program yielded important emission reductions—although they were lower than what EPA and the state originally forecast. A CARB-BAR evaluation found that by 2002, Smog Check achieved 13 to 15 percent HC emission reductions beyond those achieved by less sophisticated tests. Other researchers estimated that HC emission reductions from enhanced I/M could range from 12 to 34 percent. Additional CARB and BAR research showed that test-only facilities more consistently failed higher numbers of vehicles, and that test-and-repair stations were more likely to operate fraudulently, further reinforcing the state's migration of problem vehicles to test-only inspections (e.g., Singer and Wenzel 2003; CARB and DCA 2004; Morrow and Runkle 2005).

In 2006, the IMRC published a review of Smog Check that included some surprising and contradictory findings. In contrast to prior work, the IMRC found that by 2005, the failure rates were the same for vehicles sent to test-only and to test-and-repair stations. The findings were in contrast even with the IMRC's own work done in 2000, when they found that test-only and GPC stations (highly qualified test-and-repair stations) were superior at achieving emission reductions (CA IMRC 2000). Informal discussions with BAR staff and test-only station technicians indicated that over the years, restrictions on what constituted a California test-only station became more relaxed, perhaps accounting for the new results. The IMRC concluded, "The fundamental rationale and basis for the percentage of vehicles directed to test-only requires a reevaluation" (CA IMRC 2006, *II–13*).

Also in 2006, a BAR-sponsored evaluation found that Smog Check impacts may be short-lived. Of the 1976–1995 vehicles that had first failed and later passed Smog Check, half failed again when randomly tested on the roadside within a year. The findings were disappointingly similar to observations made more than

15 years earlier, before the California-EPA conflict (Lawson et al. 1990; Sierra Research 2006).

## The Disengagement of EPA

A 2006 evaluation by the EPA Office of Inspector General (OIG) found that from 1999 through 2005, EPA trimmed the personnel it devoted to I/M by nearly half, from 18 to 10 full-time-equivalent staff. EPA believed this to be appropriate because I/M programs had matured. OIG determined, however, that greater oversight was necessary. The 1990 CAAA mandated, "Each State shall biennially prepare a report to the Administrator which assesses the emission reductions achieved by the [enhanced I/M] program." From 1999 to 2004, however, EPA had received timely I/M reports for only 11 of 34 programs; in about 25 percent of cases, the agency never received the required reports. Nor was the problem a new one; as early as 2001, the NRC found that the majority of enhanced I/M states had failed to complete their required reports. The case of Virginia helps illustrate these outcomes. In 1999, a year after Virginia launched enhanced I/M, the state submitted to EPA an evaluation of its program. For at least the next eight years, the 1999 report stood as the last official Virginia submittal to EPA on I/M emission reductions (NRC 2001; VA DEQ 2006a; Beusse et al. 2006; Olin 2007).

The reports would not be essential if EPA had confidence that states had achieved desired outcomes. However, OIG found that states did not adequately track the fate of failing vehicles—in other words, many states had no idea what emission reductions, if any, their programs generated. Delaware, Maryland, Pennsylvania, and Virginia, for example, reported unknown outcomes for 12 to 22 percent of failing vehicles. OIG also noted wide variability in the time needed to repair failing vehicles (Beusse et al. 2006).

Even I/M implementation problems could be overlooked to a certain degree if the need for I/M had substantially diminished. The vehicle fleet gets cleaner each year, as newer cars replace older, higher-emitting ones. Thus the fleet of the early 1990s, when EPA issued its first enhanced I/M regulations, was far different from the one of the mid-2000s. Nevertheless, in the mid-2000s, states were still relying on I/M. In Maryland, for example, I/M accounted for about 40 percent of needed mobile source emission reductions. Similar findings were available for Virginia, whose officials also commented that I/M reduced air toxics not addressed by ozone or PM air quality management plans, an idea reinforced by others (M.J. Bradley & Associates Inc. 2002; Beusse et al. 2006).

Part of the reasoning behind EPA's disengagement from I/M was the agency's anticipation that by the late 2000s, the lion's share of the fleet would be equipped with OBD. An OBD-equipped fleet would, presumably, render obsolete traditional tailpipe-based I/M programs and the oversight needed to manage them. As early as 2004, about 60 to 70 percent of the light-duty fleet already was equipped with OBD. EPA's strategy, however, was built on the assumption that OBD-based I/M would successfully replace tailpipe tests.

## THE SHIFT TO OBD

### A Dodge Neon Story

On January 11, 2007, a 1996 Dodge Neon with 92,704 miles on its odometer received a Smog Check at one of California's test-only inspection facilities. BAR's HEP software had identified the vehicle as a potential high emitter, and the owner had been instructed to take it to a test-only station. The Dodge Neon, like all 1996 and later vehicles, was OBD-equipped. The Neon's OBD "check engine" dashboard light was not illuminated, and when the Smog Check station downloaded the vehicle's OBD information, the data said the vehicle was fine. In most states, the vehicle's I/M test would be finished. In California, however, the vehicle went on to receive an exhaust test, and the tailpipe told a different story: the Neon's emissions were high enough to qualify it as a gross polluter (Figure 7-2).

If the Dodge Neon's Smog Check results were an anomaly, they would be of little consequence. After all, millions of vehicles are inspected each year, and that large a population will have numerous oddities. An alternative explanation is that the first few model years of the OBD program may have had start-up difficulties as manufacturers perfected the equipment. A federal OBD work group acknowledged that model year 1996 to 1998 vehicles were the first to have more advanced OBD systems, and those vehicles would be expected to "naturally have flaws" that would be corrected in later model years (Klausmeier 2002, *54*). But unfortunately, the problem was not limited to the early introduction of OBD.

### The Overlap Issue: Differences between OBD and Exhaust Tests

During the late 1990s and early 2000s, several states found a lack of overlap between OBD and dynamometer tailpipe test results: each test identified different problem vehicles. For example, California observed that among failing vehicles receiving both an OBD and an ASM dynamometer test, 7 percent of vehicles failed both tests, 46 percent failed OBD but passed ASM, and 47 percent failed ASM but passed OBD (Amlin 2002). As of 2006, California continued to report that OBD tests missed more than three-quarters of ASM tailpipe test failures, with similar

**Smog Check Vehicle Inspection Report (VIR)**

Vehicle Information

Test Date/Time: 01/11/2007 @ 13:52

| | | | | | |
|---|---|---|---|---|---|
| Model Year: | 1996 | Make: | DODGE | Model: | NEON |
| License: | | State: | CA | VIN: | |
| Engine Size: | 2.0 L | Type: | Passenger | Transmission: | Automatic |
| GVWR: | N/A | Test Weight: | 2625 | Cylinders: | 4 |
| Odometer: | 92704 | Certification: | California | VLT Record #: | |
| Fuel Type: | Gasoline | Exhaust: | Single | Inspection Reason: | High Emitter Profile |

Overall Test Results

**YOUR VEHICLE FAILED AND EXCEEDED THE GROSS POLLUTER LIMITS**

Comprehensive Visual Inspection: PASS Functional Check: PASS Emissions Test: **GROSS POLLUTER**

Repairing your vehicle is necessary to help California reduce smog-forming emissions and reach our air quality goals.

Emission Control Systems Visual Inspection/Functional Check Results

(Visual/Functional tests are used to assist in the identification of crankcase and cold start emissions which are not measured during the ASM test)

| Result | ECS | Result | ECS | Result | ECS |
|---|---|---|---|---|---|
| Pass | PCV | N/A | Thermostatic Air Cleaner | Pass | Fuel Evaporative Controls |
| Pass | Catalytic Converter | N/A | Air Injection | Pass | MIL/Check Engine Light |
| Pass | EGR Visual | Pass | Vacuum Lines to Sensors/ Switches | Pass | Carb./Fuel Injection |
| N/A | EGR Functiona | | | Pass | Other Emission Related Components |
| Pass | Fuel Cap Functional | N/A | Ignition Timing: | | |
| Pass | Fuel Cap Visual | Pass | Wiring to Sensors | Pass | Oxygen Sensor |
| Pass | Spark Controls | N/A | Fillpipe Restrictor | Pass | Liquid Fuel Leaks |
| N/A | Fuel Evaporative Controls Functional | | | | |

ASM Emission Test Results

| | | %$CO_2$ | %$O_2$ | HC (PPM) | | | CO (%) | | | NO (PPM) | | | |
|---|---|---|---|---|---|---|---|---|---|---|---|---|---|
| Test | RPM | MEAS | MEAS | MAX | GP | MEAS | MAX | GP | MEAS | MAX | GP | MEAS | Results |
| 15 mph | 1790 | 9.5 | 0.1 | 66 | 283 | 271 | 0.57 | 2.07 | 7.99 | 487 | 1970 | 135 | **GROSS POLLUTER** |
| 25 mph | 2929 | 8.1 | 0.0 | 49 | 233 | 279 | 0.55 | 2.05 | 10.04 | 774 | 1770 | 70 | **GROSS POLLUTER** |

MAX = Maximum Allowable Emissions GP = Gross Polluter Limits MEAS = Amount Measured

Need help with your vehicle's failed smog check? You may be eligible for financial assistance to repair or retire your car. To get an application to see if you and your vehicle qualify, visit the Bureau of Automotive Repair's website at www.smogcheck.ca.gov, or call toll-free at 1-800-952-5210. YOUR APPLICATION MUST BE APPROVED BEFORE YOU RECEIVE VEHICLE REPAIR OR RETIREMENT ASSISTANCE.

Smog Check Inspection Station Information

Station Number:

Technician Name/Number:
Repair Tech Name/Number:
Software Version/EIS Number: 0403/ES819082

I certify, under penalty of perjury, under the laws of the State of California, that I performed the inspection in accordance with all bureau requirements, and that the information listed on this vehicle inspection report is true and accurate.

1-11-7
Date Technician's Signature

√ Eng light not on.

**Figure 7-2. Smog Check Vehicle Inspection Report for a 1996 Dodge Neon that Passed the OBD Test (Reported Under "Functional Check Results") but Failed the Tailpipe Test (see the "ASM Emission Test Results")**

results reported for other states, such as Illinois (Hedglin 2006; Eastern Research Group 2006).

Some lack of overlap between OBD and tailpipe test results was expected. For example, OBD continuously monitored vehicle performance, in contrast to the snapshot view of an I/M test. Thus it was easy to see in concept how OBD would identify vehicles missed by an exhaust test. What was more of a surprise, and a source of consternation among I/M professionals, was the opposite outcome: the failure of OBD systems to detect high-emitting vehicles caught by traditional exhaust tests.

Using data from Wisconsin, EPA made some of the earliest observations of what became known among air quality professionals as the overlap problem. Out of some 120,000 vehicles from model years 1996 to 1998, about 2 percent of the fleet failed either the OBD or IM240 test; however, of the roughly 1,400 vehicles failed by each test, only about 200 failed both tests. EPA observed, "What is not clear from this data is which of the two tests is more beneficial to the environment" (U.S. EPA 2000b, *10*).

Motivated by the Wisconsin findings, Colorado launched an OBD study during 2000. It found that out of about 230,000 vehicles, nearly 3,000 failed OBD, almost 400 failed IM240, and only 66 failed both tests. The state subsequently studied 109 vehicles from model years 1996 to 2000 to assess the issue. It found that the IM240 test identified high emitters 91 percent of the time, and OBD identified high emitters 53 percent of the time. Colorado concluded that when paired together, the tests found more problem vehicles than either test used alone (Barrett et al. 2005).[5]

Various reasons were found to explain why OBD and dynamometer test results differed, and why OBD failed to find high-emitting vehicles caught by traditional tests. One explanation was related to the variability of dynamometer-based inspections: results varied by test drivers as they simulated real-world driving. Most important, perhaps, was the fact that OBD tests missed emission problems when the system had not yet collected enough electronic information—a condition called "not-ready." A not-ready status could be caused willfully by battery disconnection or when someone reset the OBD monitor, a practice referred to as pre-test code clearing (Barrett et al. 2005; Eastern Research Group 2006; Hedglin 2006). However, OBD readiness problems also resulted from system errors. EPA identified readiness problems due to "design problems with the vehicle's OBD system, scan tool/inspection software problems," and other issues. Defective OBD systems affected 17 models spanning the 1996 through 2003 model years, and as late as 2008, these were the subject of further EPA investigation. For these models, EPA advised: "Until further notice, states may choose to OBD test these vehicles regardless of their readiness status *or default them to a tailpipe test*" (U.S. EPA 2008, *2, 6*; emphasis added).

Overall, experience indicated that OBD tests complemented, rather than replaced, tailpipe tests. The importance of this issue was identified early by the NRC, which stated that the lack of overlap between OBD and exhaust tests "needs to be understood and corrected" before I/M programs transitioned to solely using OBD (NRC 2001, *103*).

Nevertheless, despite the overlap issue, EPA's April 2001 OBD regulations prompted the largest single shift in I/M programs since the advent of dynamometer use in the 1990s. Once EPA mandated that states begin OBD testing on 1996 and newer vehicles, and gave states the same emission reduction credits for OBD-only testing as for the agency's original model I/M program, states abandoned traditional exhaust tests. Table 7-1 illustrates the migration of states over time to OBD-only tests. By late 2007, even California, then the only state requiring 1996 and later vehicles to pass both OBD and dynamometer tests, explored shifting to OBD-only tests to improve Smog Check's cost-effectiveness (CA IMRC 2007).

## OVERALL OBSERVATIONS ON THE U.S. I/M POLICY EXPERIENCE

Nationally, the California-EPA conflict delayed programs by many years. EPA's 1992 regulations required states to implement enhanced I/M beginning in 1995. California, after the negotiated deal and further collapse of EPA policy, began implementation in 1998; it took until 2005 for California to send more than 60 percent of tested vehicles to test-only facilities. Other states' programs were also delayed. Missouri, for example, did not implement dynamometer testing until 2000; it took Texas until 2002 to begin dynamometer testing in Dallas–Fort Worth, and until 2003 to launch dynamometer testing in Houston-Galveston. Thus, even though EPA raced to meet its 1992 court-ordered commitment, the ensuing conflict long delayed implementation.[6]

As of 2005, based on the severity of their air pollution, 33 states and the District of Columbia operated 52 separate I/M programs throughout the United States. Of these, 25 states and Washington, D.C., operated some form of enhanced I/M; about 15 of these programs were high-enhanced (Beusse et al. 2006). Following the California-EPA deal, states implemented a range of programs, many of which were less stringent than the original EPA test-only program.[7]

It is widely recognized that I/M programs did not achieve the emission reductions forecast in the early 1990s. A study by 12 of the world's major automotive and petroleum-related manufacturing companies observed that although detecting high-emitting vehicles, in principle, should "not be a great challenge," the I/M experience had been "decidedly mixed." They noted that "even when vehicles are determined to have malfunctioning emissions control systems, authorities sometimes are reluctant to force owners to bring these systems into compliance" (WBCSD 2004, *100*).

As of 2007, EPA continued to cite I/M as one of the most cost-effective mobile source control options available (U.S. EPA 2007).[8] A 2007 California evaluation found, however, that program costs varied widely by state, depending on test type. Among states that charged for testing, motorists paid from $16 for an idle test in Tennessee's centralized program to $86 for an ASM dynamometer

| *Year* | *Tailpipe-only* | *Tailpipe and advisory OBD* | *Tailpipe and mandatory OBD* | *OBD-only* |
|---|---|---|---|---|
| 1999 (pre-OBD regulations) | Arizona<br>California<br>Colorado<br>Connecticut<br>Delaware<br>Illinois<br>Indiana<br>Maryland<br>Massachusetts<br>Missouri<br>New York<br>Ohio<br>Pennsylvania<br>Rhode Island<br>Texas<br>Virginia<br>Wash., D.C.<br>Wisconsin | | | |
| 2003 | Maryland | Colorado<br>Connecticut<br>Massachusetts<br>New York<br>Ohio[a]<br>Rhode Island<br>Virginia | California<br>Delaware[b]<br>Illinois[a]<br>Pennsylvania[a] | Arizona<br>Indiana<br>Missouri<br>Texas<br>Wash., D.C.<br>Wisconsin |
| 2007 | | Colorado | California | Arizona<br>Connecticut<br>Delaware<br>Illinois<br>Indiana[c]<br>Maryland<br>Massachusetts<br>Missouri<br>New York<br>Pennsylvania<br>Rhode Island<br>Texas<br>Virginia<br>Wash., D.C.<br>Wisconsin |

**Table 7-1. Impact of EPA's 2001 OBD Regulations on the Evolution of Tests for 1996 and Newer Light-duty Vehicles**

*Notes*: Table includes high-enhanced I/M programs as of 2003. Advisory OBD means vehicles were not failed solely due to the OBD test

[a] Cincinnati discontinued I/M after 2005; Illinois OBD tests were phasing in; Pennsylvania OBD tests were being piloted

[b] Before 2005, Delaware vehicles failing OBD twice were certified if they passed an exhaust test; after 2005, the test became OBD-only

[c] After December 31, 2006, I/M was discontinued in southern Indiana but continued in northern Indiana

*Sources*: U.S. EPA 1999, 2003b; state websites; personal communications with state officials

test in New Jersey's decentralized program. California motorists paid $55 to $65 per test, and the Smog Check program's total costs (including inspection fees, vehicle repairs, and state-funded assistance to consumers) exceeded $830 million (CA IMRC 2007).[9]

Also by 2007, U.S. I/M programs were on the cusp of a dramatic shift. A key question was whether and for how long states would continue dynamometer testing. Dynamometer equipment cost upward of $40,000 to purchase in the mid-2000s. Fewer inspections required dynamometer use, however, as pre-1996 (non-OBD) vehicles disappeared. Consequently, I/M programs deemphasized pre-1996 cars. For example, effective in 2007, Missouri decided to forgo all emission testing of pre-1996 vehicles (MO DNR 2005).

But before disappearing from the fleet, pre-1996 vehicles remained an important residual contributor to emissions. EPA staff estimated that in 2012, for example, pre-1996 vehicles would constitute less than 15 percent of registered vehicles but contribute more than 40 percent of fleet HC emissions (Tierney 2007). Thus, in the wake of the flawed movement to OBD-only tests, states struggled with the ongoing emission ramifications of the pre-1996 vehicle fleet. In California, as the state's IMRC promoted a move to OBD-only tests, the review committee pondered the implications: "The related problem is how to keep tailpipe testing alive and viable for the next decade" (CA IMRC 2007, *38*).

In 2007, as California updated its air quality management plans, the state continued to rely on Smog Check to meet its vehicle-related emission reduction goals. California planned to improve Smog Check by requiring, among other things, improved evaporative emissions tests, more stringent pass–fail thresholds, and annual instead of biennial inspections for older and high-mileage vehicles (CARB 2007). As illustrated by California's actions, despite I/M's checkered history, it remained, well into the 2000s, one of the few viable approaches to find and control at least some fraction of high-emitting vehicles.

One can certainly point to both good and bad elements in Smog Check's legacy. On the positive side, the California-EPA pilot program addressed important technical questions. Following the California-EPA deal, new regulatory freedom fostered the tailoring of state programs to local interests. Over the long term, state experimentation and evolving insight led, in some cases, to improved programs. On the negative side, however, the conflict reinforced public discontent over Clean Air Act programs targeting individuals rather than corporations and nearly resulted in the reopening of the 1990 CAAA. The political maelstrom that engulfed EPA in the early 1990s was followed over time by the agency's reticence to engage in state oversight, lax program implementation on the part of some states, and EPA's premature promotion of OBD-based tests. Overall, the conflict delayed and reduced environmental benefits.

## NOTES

[1] Some of these same findings were anticipated by Marc Pitchford, who worked at an EPA laboratory in Las Vegas, and his University of Nevada colleague Brian Johnson. In 1993, they published an article that discussed RSD's limitations as well as its potential use for I/M. They advocated regularly scheduled conventional and RSD-based I/M tests, plus RSD-based random monitoring (Pitchford and Johnson 1993). Although Pitchford worked for EPA, his Nevada location apparently separated him from Ann Arbor decisionmakers; note that, as discussed in Chapter 3, North Carolina–based EPA scientist Ken Knapp evaluated RSD as well but was also disconnected from the Ann Arbor–based policy process.

[2] The ASM test identified about the same amount of problem emissions as the IM240 test. EPA had questioned whether ASM could perform well without falsely failing too many clean vehicles. For 1981 and newer vehicles, the pilot study found the false failure rate was nearly the same for both tests. A key question, however, was whether vehicles repaired to meet the ASM test would reduce as much pollution as vehicles repaired to meet the IM240 test. The pilot study found that IM240-based repairs were slightly better at reducing 80 percent of excess emissions. To reduce 90 percent of excess emissions, virtually the same number of repairs had to occur using either test (see Klausmeier et al. 1995).

[3] The U.S. Agency for International Development also found "For developing countries, experts are virtually unanimous in recommending test-only facilities for I/M programs. This conclusion is based on the poor performance of the test-and-repair approach in developing countries, and the far-from-perfect record in industrialized countries as well" (U.S. AID 2004, *22*).

[4] As a result of the California-EPA deal, the state had committed to send 15 percent of vehicles to test-only stations, a commitment it met by 1999 (three years after the deal-imposed deadline). Later, to improve performance once Smog Check proved less effective than hoped, the state increased the test-only requirement to 36 percent of the fleet. The method used to define 36 percent of the fleet proved to have an interesting outcome. The state defined all 1976 and newer gasoline vehicles as the I/M fleet. By 2005, however, vehicles six years old and newer were exempt from Smog Check. Thus at this point, the state directed to test-only about 48 percent of vehicles actually subject to testing. In addition, 40 to 50 percent of motorists voluntarily sought test-only inspections (Klausmeier et al. 2000; Weisser 2006). Therefore, by 2005, more than 60 percent of vehicles subject to Smog Check received test-only tests.

[5] EPA also laboratory-tested 153 vehicles from Colorado, Arizona, and Michigan and found higher overlap between OBD and exhaust results (Gardetto et al. 2005). However, an independent analysis of the EPA study found that "in an I/M setting, an OBD-only inspection-and-repair program will worsen air quality over the near term rather than improve it, given that the highest emitters are missed by OBD, while OBD identifies many marginal and low emitters" (Lawson 2005). Further analysis of the same data found that 49 percent of emission reductions came from repairing just three vehicles, one of which failed only the IM240 test. The analysts found that failures missed by OBD "resulted in significant loss" of emission reductions (Lawson and Gardetto 2006; Stedman 2006).

[6] The premise behind EPA's 1992 I/M policy was well understood for many years prior to the policy's release. As early as 1979, an article pointed out that centralized I/M programs entailed "less administrative, auditing, and surveillance costs," and that a weakness of garage-based programs was "the potential for fraud; close governmental scrutiny is necessary." It also noted that "because it simulates actual driving conditions, the loaded mode [dynamometer] test provides a better indication of actual driving conditions than does the idle mode test" (Reitze 1979).

[7] A year 2007 scenario illustrates the program diversity that emerged after the Smog Check debate. Assume a 1994 station wagon and a 2000 sedan located in each state. In about a third of states,

pollution is minor and the vehicles' emissions are not tested. In about half of states that test vehicles, the cars are subject to high-enhanced I/M. In almost all high-enhanced programs, the 2000 sedan receives only an OBD test; the 1994 wagon receives only a dynamometer test. In Missouri and Illinois, the 1994 wagon is exempt as too old to test. Among states that test, about half have test-only and half have test-and-repair programs; California has both systems. Some states test vehicles annually; most test every two years. Four states also deploy RSD at the roadside: Colorado uses RSD to exempt low emitters from I/M; Texas and Virginia use it to catch high emitters; and in California, owners of high emitters found by RSD might be offered money to repair or scrap their cars.

[8] An important gap in the pollution problems addressed by U.S. I/M programs was that tests did not measure PM. Fine PM, however, is an important contributor to health risks from air pollution, and research has shown since the 1990s that gasoline-powered vehicles can contribute a large fraction of fine PM emissions (Watson et al. 1998).

[9] Test costs across states were adjusted by the IMRC to reflect California's cost of living.

# PART III
# ANALYSIS

CHAPTER 8

# WHY THE CONFLICT OCCURRED

Given the problems that resulted from the Smog Check debate, an obvious question is, why did the conflict occur? This chapter addresses that question in two ways. It begins by exploring the debate's root causes, then takes a broader view and places the conflict in the context of how federal agencies operate, drawing insights from the policy analysis literature.

## ROOT CAUSES OF THE CONFLICT

Although many factors contributed to the conflict, three were of overriding importance: EPA's inability to build a scientific consensus on how to improve I/M programs, use of prescriptive command-and-control regulations, and failure to bargain in the political arena. Circumstances also played a role—the economic recession and the transition between the outgoing Bush and incoming Clinton administrations contributed to the difficulties that led to the conflict.

## The Lack of a Technical Consensus

"The first battleground in any environmental controversy is the scientific depiction of cause and effect," noted Judith Layzer in her introduction to *The Environmental Case* (2002, 6). The job of finding and fixing high-emitting vehicles would have been much easier if the major policy participants could have agreed on the best approach to the problem. During the Smog Check debate, however, there was no consensus view about the right course of action. As EPA advocated on behalf of its regulations, state officials, scientists, environmentalists, I/M service station owners, and others promoted alternative views. Troubled over competing technical claims that muddied the I/M debate, the California legislature commissioned the RAND Corporation to assess the technical merits of EPA's policy. RAND's findings, released in August 1993, precipitated the collapse of EPA's policy: "Based on our analysis of available data, we believe that the California SMOG CHECK system must be restructured. . . . Based on effectiveness in reducing emissions, we find no empirical evidence to require [as EPA did] the separation of test and repair" (Aroesty et al. 1993).

As the RAND findings made clear, EPA failed to make its technical case in the political arena. Not only did the agency fail to make its case, but it was perceived as being technically wrong. Legislators and others resented EPA for attempting, as they perceived it, to force a solution on California simply because the federal government thought it had the authority to do so, rather than because the federal government's solution made sense.

Data collected after the debate reinforced the idea that EPA had erred technically prior to the conflict. The California-EPA pilot study results showed that the agency's program failed to prove as superior as predicted, and implementation problems occurred in states such as Maine. Why had EPA erred technically? What was the nature of the agency's policy analysis process that impeded its ability to better forecast the technical merit of its proposal? Answers to these questions reveal important systemic issues with widespread implications for EPA's environmental management efforts.

EPA, during the I/M rulemaking process, bundled together its best scientific evidence to support the case for test-only I/M programs. This evidence, however, reflected the shortcomings of the regulatory process. Sheila Jasanoff, in her work *The Fifth Branch: Science Advisers as Policymakers* (1990), noted important differences between "research science" and "regulatory science." In the research arena, scientists can take longer to derive information they want; in the regulatory arena, waiting for more data is synonymous with a decision not to act. Thus regulatory science, or the building of a scientific case for regulatory decisionmaking (one might call it "science in a hurry"), generally meets a lower standard of proof than scientific claims put forth in the peer-review literature (Jasanoff 1990). As policy analyst Mark Powell observed in *Science at EPA* (1999), EPA is not a science agency, but a regulatory agency; the agency's scientific work therefore is tailored to meet the shorter-term demands of its regulatory responsibilities.

Although policymakers and stakeholders generally try to access as much scientific information as possible, the regulatory process rarely facilitates completing new research. Regulatory actions thus reflect whatever scientific information is available at the time of a decision and usually occur in an atmosphere of scientific uncertainty. As noted by others, this uncertainty provides flexibility in framing the policy problem, depending on which group is articulating its position (Layzer 2002).

When the science is uncertain and the time available to make decisions is compressed, criteria other than scientific fact become paramount. As Powell observed, EPA makes decisions based on a mix of legal, political, economic, and scientific concerns. Science is not the main decisionmaking component. Powell's work, a study on the use of scientific information by EPA, offers insights directly relevant to the I/M debate:

- Policymakers are, typically, attorneys with no formal scientific training; communication between agency scientists and decisionmakers is often poor.
- Appointed leaders with relatively short tenures focus on short-term goals, rather than longer-term research and development.
- Besieged by a workload that exceeds its resources, EPA uses "simplified science" for decisionmaking, often triggering later controversy.
- High-stakes decisions can mean that politics and economics overwhelm science as decisionmaking criteria. (1999, *3–5, 112–113, 122*)

Given the limitations of available data, Powell found that "much of what passes for 'environmental science' is, in fact, a negotiated consensus among experts" (1999, *127*). Jasanoff, among others, offers similar insights: "Regulatory practices ... support the thesis that negotiation—among scientists as well as between scientists and the lay public—is one of the keys to the success of the [scientific] advisory process" (1990, *234*). Negotiating scientific positions takes time; in the Smog Check situation, however, EPA was forced by court order to quickly prepare regulations.

Powell concluded that EPA is most likely to use scientific information when the agency's leadership pushes for it, when there is peer review of its decisions, when external pressure to consider scientific evidence is brought to bear, when EPA has adequate scientific resources and time, and when the relevant statutory provisions require or enable scientific inquiry. Many of these conditions were not met during the period leading up to the Smog Check conflict; the EPA's leadership was in transition between the Bush and Clinton administrations, there was no scientific peer review of potential I/M policy options, and the 1990 CAAA deadlines left the agency little time to fulfill its regulatory responsibilities.

Writing before the I/M conflict, Jasanoff offered further insights about the notice-and-comment rulemaking process that presaged outcomes like those observed in Smog Check. Using case studies of another technically complex environmental problem—cancer risk guideline development—she showed that the regulatory process handles new scientific information best when EPA first builds a scientific consensus among experts external to the agency. Jasanoff described, as an example, EPA's assessment of the risks from formaldehyde exposure. The assessment process ran from 1979 through 1987 and used a succession of scientific workshops, external peer reviews, and science advisory boards to examine and interpret data. The process allowed the agency to arrive at a final regulatory decision that gained acceptance in the scientific community. The key to this successful outcome was that EPA separated the building of a scientific case from the notice-and-comment rulemaking process. However, the process took eight years (Jasanoff 1990). Although the eight-year formaldehyde assessment process may be an extreme example of the time needed to build scientific consensus, the court-constrained I/M rulemaking schedule fell at the other end of the time-available spectrum.

Jasanoff also observed that the notice-and-comment rulemaking process is by nature an adversarial process, in which scientists function "as just another interest group." Engagement of the scientific community prior to rulemaking allows EPA to build its scientific case before the more adversarial regulatory process unfolds (Jasanoff 1990). The importance of building a technical consensus has been illustrated by others as well. Layzer, for example, touched on the importance of technical consensus building and its relationship to other policy-setting criteria such as costs and benefits (Layzer 2002). To engage in consensus building prior to rule development, agencies need sufficient foresight to establish and build consensus over a research agenda; collect, share with stakeholders, and evaluate data; and build consensus on findings. These steps take time and resources.

In addition, Jasanoff cautioned that the rulemaking process can "freeze" the science by locking the government into a particular policy solution. Specifying mandates in too much detail can be problematic in situations where information is emerging that is shaping or improving the understanding of a problem and its solutions (Jasanoff 1990). With Smog Check, the release of EPA's prescriptive mandates was juxtaposed with a scientific state of flux over I/M program options, as exemplified by the emergence of RSD and the lack of needed data to compare test-only with test-and-repair systems.

These brief observations from the literature are not meant to serve as a comprehensive critique on the use of science at EPA. Rather, they highlight long-standing examples, relevant to Smog Check, of the difficulties the agency has encountered in obtaining and applying scientific information.

As these observations help indicate, EPA's formulation of a national I/M policy was scientifically handicapped in four important ways. First, the agency failed to anticipate the need for a scientifically defensible evaluation of I/M implementation experiences. Inhibited by the lack of a long-term, science-based research agenda, EPA was forced—by Congress and a court-ordered deadline—to assemble a scientific record, formulate a national I/M policy, and develop implementation guidance for states, all within two years of the 1990 CAAA becoming law. Thus the agency's model I/M program rested on a weak foundation of supporting information.

Second, EPA issued its regulations without initially building a technical consensus within the scientific community about the best approaches for identifying and repairing problem vehicles. Ideally, the agency should have sought technical input and peer review prior to entering the notice-and-comment stage of developing a national I/M policy. Such a process could have helped EPA better understand and respond to the technical shortcomings of available information, identify information needs, facilitate new data gathering, and solidify technical support for a practical solution. The abridged, statutory, and court-ordered schedule made such a task difficult.[1]

Third, the structure of the final I/M regulations left states little room for innovation or experimentation. The regulations, to use Jasanoff's term, "froze" the science, offering scant opportunity to consider the role of emerging technology, such as remote sensing.

Finally, the scientific knowledge chasm that separated the technical career staff at EPA from the administration's senior political leadership—a situation Powell found in other EPA cases—made it difficult for the political appointees to select the right decisionmaking path. As described in the earlier chapters on the Smog Check conflict, White House staff in particular, far removed from the technical details surrounding I/M, encouraged EPA to improve communication of the agency's messages regarding the benefits of test-only programs. Trusting that EPA's career staff had done its technical homework, the appointees focused on the outreach portion of the I/M problem, including relationships with the press, public interest groups, and elected officials. (Chapter 9 discusses the roles of the press and the public in environmental policy setting.)

## EPA's Use of Command-and-Control Regulations

At the time of the Smog Check controversy, EPA's culture promoted the use of command-and-control regulations—rules that often narrowly defined the specific control actions and technologies to be employed to reduce pollution. By the early 1990s, command-and-control regulations had already achieved substantial (though incomplete) progress toward clean air, but as evidenced by Smog Check, they could engender a negative response from those being regulated. California arguably had been leading the nation and the international community in addressing air pollution problems for more than 30 years when the debate arose. EPA's prescriptive I/M mandates, paired with a lack of technical data to support them, sparked a nonpartisan backlash among California officials who believed philosophically that states—and especially California—should have the freedom to solve their own problems. Why, after years of previous federal command-and-control regulatory actions, did the I/M situation produce such a strong backlash? The answer to that question lies in the historical context for the debate period: EPA's 1992 I/M rule ran counter to growing interest in a new regulatory paradigm.

During the 1980s and early 1990s, there was a call to promote more innovative, results-based government programs. In 1982, U.S. policy analyst John Naisbitt published the book *Megatrends*, in which he noted that the United States had been experiencing a movement since at least the 1960s toward greater decentralization of governmental authority and responsibility. He called the movement "a new assertiveness" among states and predicted that "governors and state legislatures are gearing up for new states' rights battles" (Naisbitt 1982, *104–108*).

Ten years later, another seminal work emerged: *Reinventing Government* (Osborne and Gaebler 1992), which advocated moving away from government "business as usual" toward alternative policy-setting approaches that focused on results and customer service and fostered an entrepreneurial spirit within the public sector. As a *New York Times* book review noted on March 8, 1992, *Reinventing Government* had "already found its way onto the night stands of Presidential candidates like Bill Clinton," and by October 1992, just before EPA issued its final I/M regulations, the book had become so popular it was already in its eighth printing. On March 3, 1993, less than two months after taking office, President Clinton announced the formation of a National Performance Review (NPR) task force, chaired by Vice President Al Gore, "to redesign, to reinvent, to reinvigorate the entire National Government" (NPRG 1993). When the task force released its findings six months later, Osborne and Gaebler's *Reinventing Government* provided the template for the Clinton–Gore reform effort (Frederickson 1994; Kettl 1998).

*Reinventing Government* noted that the traditional approach to government regulation is through command-and-control actions: "lay down rules and order people to comply." Osborne and Gaebler cited EPA as a "perfect example" of a federal bureaucracy that employed these traditional techniques: "Ever since it created the Environmental Protection Agency (EPA), the federal government has relied primarily on a command-and-control strategy." The authors discussed the drawbacks with command-and-control regulations, pointing out that such a strategy "relies on the threat of penalties—but in a political environment, many of those penalties can never be assessed." They also observed that "regulations that specify the exact technology industry must use to control pollution discourage technological innovation" and that "the command and control approach slaps the same requirements on industries all over the country" and applies an expensive "one-size-fits-all" approach (Osborne and Gaebler 1992, *299–301*).

Other researchers have noted that by the 1990s, a new regulatory environment existed. For example, in a review of the policy implementation literature, Michael Hill and Peter Hupe showed how the 1980s and 1990s movement away from command-and-control-based regulatory approaches reflected the ideological tenor of the times. They observed that by the 1990s, distrust in government was growing because of policy failures following what they termed the 1950s–1970s big-government, "great expectations" era. The result was a movement to emphasize management efficiency in government, rather than social values (Hill and Hupe 2002).

Unfortunately, the late 1980s and early 1990s was an awkward period for U.S. environmental policy reinvention. At the same time that states sought greater autonomy from the federal government, the U.S. Congress was taking a more activist role in shaping environmental policy. Congressional actions were motivated, in part, by the controversial tenures of EPA administrator Anne Gorsuch Burford and U.S. Department of the Interior secretary James Watt, both appointees of Ronald Reagan, who served as president from 1981 to 1989. Burford and Watt reflected Reagan administration views characterized as mistrustful and "openly skeptical of the historical missions" of federal environmental and natural resource agencies (Auer 2008, *76*; see also Kraft and Vig 2003). Although the 1989 transition from Reagan to George H.W. Bush marked the beginning of greater "collegial relations with [federal agency] careerists and respect for their neutral competence" (Auer 2008, *75*), Congress remained motivated in the late 1980s to prescribe more rigorous environmental mandates, especially regarding clean air. The 1977 CAAA failed to result in states achieving clean air by the act's 1987 deadline, and as a result, the 1990 CAAA, crafted just following the Reagan administration and signed into law by George H.W. Bush, included far greater specificity than the 1977 legislation it replaced. In addition, many states, by failing to adequately implement required environmental controls, helped motivate Congress to provide a more rigorous clean air statute (Portney 1990; Reitze 1996; Kraft and Vig 2003).

As others have noted, with the 1990 CAAA, Congress set overly ambitious statutory schedules and did not endow EPA with sufficient resources to meet the law's many mandates. The result was EPA's frequent inability to meet statutory deadlines (e.g., Portney 1990). Consequently, the agency was targeted by lawsuits to compel it to meet its statutory obligations—a situation exemplified by EPA's court-ordered publication of its 1992 I/M regulations. The I/M case study therefore is but one of many examples where EPA was under severe deadline pressure to meet its CAAA obligations. Congress charged EPA to meet 538 separate requirements under the first several portions of the 1990 CAAA. By February 2000, EPA had completed 409 of these requirements. However, well over half of these requirements had statutory deadlines, and EPA missed 80 percent of these milestones (U.S. GAO 2000).

Thus, at the very time when there was a growing demand for state autonomy, a deemphasis of traditional command-and-control federal rulemaking, and a call for regulatory innovation, EPA was under tremendous pressure simply to stay abreast of its many new statutory obligations. In addition to statutory and judicial pressure, EPA was under pressure to address heightened interest in automotive pollution. Career staff crafted the I/M policy at a time when air pollution experts were absorbing findings from tunnel studies that showed automotive emissions were a far more important pollution source than previously had been acknowledged. Challenged by competing demands, EPA managers were hard-pressed to conduct regulatory business as usual, let alone find the resources and time to innovate and attempt new regulatory approaches. Instead, the management team typically relied on its standard regulatory methods to address the many milestones included in the 1990 CAAA, including those relating to enhanced I/M.

## EPA's Failure to Bargain in the Political Arena

Vehicle inspection programs rely on the successful participation of millions of everyday motorists and their vehicle repair technicians. Any program that effectively involves the vast majority of the public naturally garners the interest of elected officials who serve at their pleasure. In such circumstances, program design and implementation need to pass public and political acceptability tests to avoid noncompliance or, in more challenging cases, political intervention to halt or modify the program. Unfortunately, EPA's model I/M program failed to pass those tests, as evidenced by comments from California state senator Steve Peace, who spoke on behalf of the legislation that enacted the California-EPA compromise: "the federal government's proposal was an absolutely unenforceable, unworkable, impossible to achieve concept ... something that clearly could not work with the millions of people and the millions of cars that we would have had to have dealt with" (Peace 1994).

Testing for a program's political acceptability is a basic tenet of effective policy analysis and development. As policy analyst Eugene Bardach observed in what he described as the "assemble some evidence" stage of a good policy analysis, "The process of assembling evidence inevitably has a political as well as a purely analytical purpose," and therefore involves touching bases and gaining credibility with the policy stakeholders, and brokering consensus among them. Similarly, he observed that policy options must be judged using "practical criteria" such as political feasibility, a concept observed by others as well (Bardach 1996, *17*, *31*). The importance of politics is emphasized by an apt quote from constitutional theorists Anthony Bertelli and Laurence Lynn, who noted that "though *policy implementation* is technical and specialized, *policy making* is and should be political" (2006, *98*).

A key distinction about the I/M situation, in contrast to many other regulations promulgated by EPA, is that the I/M rule necessitated the active involvement of elected officials to pass authorizing legislation for new inspection programs. Legislative bodies are inherently designed to promote bargaining and political compromise. EPA, however, established its I/M policy through a traditional notice-and-comment rulemaking process.[2] The rule development process did involve soliciting and responding to approximately 300 public comments concerning the draft I/M rules (U.S. EPA 1992e). The comment process focused, however, on seeking input from the professional air quality community; not many members of the general public, after all, scan proposed regulations in the *Federal Register*.

Once the I/M regulations were finalized, EPA worked to persuade California officials of the merits of a test-only program, rather than to negotiate a compromise program acceptable to the state. These efforts peaked in the early months of the Clinton administration, when EPA threatened to sanction California unless it accepted the agency's I/M plan. EPA's threatened use of sanctions, without a corresponding offer to compromise, politically isolated the agency. The sanctions threat enhanced bipartisan antagonism against EPA among California's elected officials. Overall, the agency failed to recognize what was within the range of a politically acceptable solution in California.

## Other Factors: The Recession and the Bush–Clinton Transition

At the time of the Smog Check debate, the United States—particularly California—was in the midst of an extended recession. California began 1990 with a relatively low, 5.0 percent unemployment rate; however, the state's economic situation deteriorated substantially over the next three years. California unemployment peaked at 9.7 percent early in 1993, just as the state and EPA diverged over I/M policy (U.S. unemployment peaked at 7.8 percent in mid-1992) (CEDD 2003; BLS 2003). EPA promoted taking inspection business away from service stations and giving it to test-only facilities. Despite the agency's assertions that its regulations would increase repair work and jobs, the recession was one of the many factors that motivated state lawmakers to find an alternative solution to the vehicle inspection problem.

EPA finalized its I/M policy in November 1992, during the last months of the Bush administration. The Bush-to-Clinton transition meant that key policy decisions were made before the Clinton-EPA team was in place. Yet it fell to Carol Browner and her staff to implement these policies and resolve the problems that resulted.

## EPA OPERATING PRACTICES THAT AFFECTED THE DEBATE OUTCOME

On October 4, 1993, Clinton presidential appointee Felicia Marcus became the EPA regional administrator for the southwestern United States. At that point, EPA and California had been at odds over I/M for nearly a year. The agency had already threatened the state with sanctions for failing to produce approvable I/M legislation and only recently had begun to yield to political pressure to soften its inflexible policy stance. A lawyer and political activist who had worked on air and water pollution issues in Southern California, Marcus stepped into the I/M debate at the critical juncture when EPA was preparing its regulatory package to announce California sanctions. Just nine days after being sworn into her new post, Marcus was briefed by EPA career staff about the I/M conflict that had flared between California and the federal government. "Why can't we just do what State Assemblyman Katz is saying and modify the current program?" asked Marcus, alluding to the alternative I/M proposals advocated by leaders in the California legislature.

This situation involving Felicia Marcus points to a systemic challenge in running bureaucracies. Political scientist Theodore Lowi once remarked that in U.S. bureaucracies, "there is a constant tension between elected officials and career administrators" (Lowi 1976, *456–457*). Marcus, arriving at Smog Check with a fresh perspective, challenged the fundamental premise of the career staff's approach to I/M.

Lowi and many others have offered insights into how government works. The discussion that follows takes a broader view of the I/M case study by using examples from some of the policy literature's classic texts to help examine the unfolding of the California-EPA debate, drawing parallels between many of the Smog Check episode's difficulties and others' characterizations of the standard operating practices of federal bureaucracies. The material is organized in chronological order by publication date.

## Lindblom and the Limits of Incremental Decisionmaking

In a 1959 article titled "The Science of 'Muddling Through,'" Charles Lindblom described how policymakers usually took only incremental decisionmaking steps toward problem-solving and rarely ventured forward with dramatic policy leaps. The resulting policies, said Lindblom, are not comprehensive, fail to consider all outcomes, and thus usually are only partly successful and often can have unintended consequences. Given the uncertain outcomes that stem from any policy decision, Lindblom offered that the test of whether a policy is "good" or not is whether the "contestants" in the policy-setting process agreed on an outcome. The policy process itself, according to Lindblom's insights, is an iterative effort: "Policy is not made once and for all ... it is made and remade endlessly" (Lindblom 1959, *83–84, 86*).

This perspective, which does not tolerate the sudden emergence of new technology or information, anticipates the problems encountered during the I/M debate. Lindblom asserted that policymakers usually consider implementation alternatives that vary only modestly from the existing world as it is known. An obvious extension of Lindblom's thinking to the I/M debate is the sudden emergence of remote sensing technology as a new method for detecting problem vehicles. In Lindblom's policymaking model, EPA was not prepared to embrace RSD as a replacement for or serious complement to traditional tailpipe testing; the RSD concept was simply too far from the existing testing paradigm. Also, Lindblom's model may be extended to show that EPA's failure to achieve political consensus for its model-enhanced I/M program doomed the policy from the outset. Lindblom argued that the ultimate test of good policy is its acceptance by the policy contestants, a threshold test EPA failed to meet.

Finally, Lindblom's model offers insights that extend beyond the root causes of the conflict to an understanding of how I/M policy evolved over time. Lindblom accurately predicted the likelihood of unintended policy consequences. The various pilot programs and full-scale implementation efforts that followed the California-EPA negotiations bore testimony to the fact that both the agency and the state experienced unanticipated real-world implementation problems. Lindblom's insight that policies continue to evolve also provides an accurate summation of the I/M policy experience. Far from setting policy in stone, the March 1994 California-EPA agreements marked only the beginning stages of I/M policy evolution. As discussed earlier in this book, state and federal I/M policies shifted substantially in the following years.

## Downs and the Role of Personal Staff to Appointed Officials

The 1967 publication *Inside Bureaucracy*, by Anthony Downs, provided a government behavior model with two insights especially relevant to the I/M case study. First, it included several paradigms to explain individual and organizational behaviors. The individual behavior models were premised on the idea that self-interested individuals shape decisionmaking within an organization, a process the author termed "biased behavior." Downs also noted the behavior bias of high-level agency staff. High-level officials, said Downs, typically adopt a "wait and see" attitude toward the many perceived crises that are called to their attention. Given the tendency of individuals to exaggerate, Downs continued, the "threats" that are raised to senior managers usually fail to materialize. Hence leaders have a "healthy skepticism" about whether problems are truly important. Unfortunately, this also means that high-level officials tend to be insensitive to impending crises. When a real crisis emerges, said Downs, it may be after the fact before the officials recognize and respond to the crisis. Downs combined his models for individual behavior with a broader view of agency behavior. He described the "rigidity cycle" that affects agencies that have experienced extensive growth over time and said that they are frequently incapable of fast or novel action (Downs 1967).

Second, Downs identified the advantages to appointed leaders of having their own personal staff to advise them, separate from others. Major operational officials "gradually develop staffs to give them technically specialized advice." Downs observed that "a top-level official can still derive great benefits from a separate staff. A large staff can function as a control mechanism 'external' to the line hierarchy, promote change in opposition to the line's inertia, and act as a scapegoat deflecting hostility from its boss" (Downs 1967, *153–154*).

During the early stages of the I/M debate, as the divide between California and EPA began to widen and the Clinton administration first took office, EPA administrator Carol Browner lacked the infrastructure of her own key policy advisers to provide independent guidance and analysis of the I/M situation. The California-EPA I/M conflict gathered momentum throughout the first eight months of 1993, yet it was not until late summer that President Clinton appointed Mary Nichols as the EPA assistant administrator for air quality issues, and not until October that Felicia Marcus was sworn in as the presidential appointee to head the southwestern EPA office. According to Downs, given the many crises that envelop a senior official such as Browner, it would have been difficult for her at the outset of the conflict—lacking the independent counsel of personal staff such as Nichols and Marcus—to perceive the relative importance of the developing I/M policy crisis.

Combined with the handicap of lacking a core appointed team, Browner faced the difficulties of managing an agency that had grown more than fourfold over its 22-year history. When first created on December 2, 1970, EPA had a workforce of 4,000 and an annual budget of $1 billion. By 1992, when Browner took office, EPA's workforce had grown to 17,000 and its annual budget to $6.7 billion (U.S. EPA 2003a). With the changes, as Downs predicted, came a slowing down of the agency's ability to respond to problems. During a 1993 oral history interview, Alvin Alm, an assistant administrator under EPA's first leader, William Ruckelshaus, was asked, "What are the most significant challenges facing the agency in the 1990s, substantively, politically, and managerially?" Alm replied, "Well, one of the biggest challenges is to speed up agency processes. Promulgating a regulation now takes over four years, and sometimes up to eight years" (U.S. EPA 1994b, *22*). Thus Browner's career team was challenged to rapidly respond to new events as they unfolded.

## Allison and the Importance of Standard Operating Procedures and Politics

In *Essence of Decision*, an examination of the early 1960s Cuban Missile Crisis, Graham Allison identified several government decisionmaking models. As Allison explained, government behavior is not necessarily the result of a monolithic, rational actor ("Model I" in Allison parlance); rather, government "consists of a conglomerate of semi-feudal, loosely allied organizations, each with a substantial life of its own." Allison observed that government decisionmaking is best explained as groups of actors following standard operating procedures, or SOPs ("Model II"), and as the outcome of political bargaining that takes place within large organizations ("Model III"). He also suggested that organizations have "tendencies" in the way they approach policy problems, meaning an agency culture that helps govern staff actions (Allison 1971, *67–68*).[3]

Allison's models of government decisionmaking draw a distinction between the appointees who assume agency leadership positions within each administration and the career staff that transcend administrations. In many cases, career staff define the policy options available to appointed decisionmakers. Elected leaders and their appointees do not assume command of a "clean slate." They inherit previous decisions, analyses, and efforts to winnow policy options down to a few coherent opportunities. Thus Allison viewed large policy acts less as the result of a bold decision by a single actor and more as the cumulative output of "innumerable and often conflicting smaller actions by individuals at various levels of bureaucratic organizations." Politics within a government agency play a prominent role, according to Allison. He portrayed policy outcomes as the result of "compromise, conflict, and confusion of officials with diverse interests and unequal influence" (Allison 1971, *6, 78–79, 162*).

These insights help illustrate how EPA's Office of Mobile Sources (OMS) played such a domineering role in the I/M debate and underscore why EPA's model I/M program emerged in the form that it did. As Allison's models would predict, the tendency of OMS, and EPA in general, to issue command-and-control regulations precluded EPA from seriously entertaining alternative I/M approaches. Because of the "semi-feudal" nature of each EPA office, there was little impact on decisionmaking at OMS, for example, when EPA scientists in North Carolina and Nevada disagreed with OMS concerning the merits of RSD (e.g., Pitchford and Johnson 1993). Separation of these scientists from Michigan-based OMS, institutionally and geographically, prevented peer input that might have averted the bitter RSD debates that spilled into the press and made consensus difficult to achieve (e.g., see Spencer 1992).

More important, as Allison's model predicts, the I/M policy options presented to newly appointed Administrator Browner in early 1993 were already narrowed by the actions of career staff and the I/M rulemakings that preceded her tenure. The I/M regulations had been finalized the previous November. Then, as EPA fielded California's questions about how to interpret them, the agency solidified its positions regarding state options. EPA's rigid stance on its model I/M program became firmly established on January 26, 1993, when the agency sent its first formal comments to California (Howekamp 1993). The comment letter sent on this date had been under development for weeks before its release; when it was signed, President Clinton had been in office a mere six days. By the time Carol Browner took the helm at EPA, many of the agency's key I/M policy decisions already had been made.

Allison emphasized the importance of intra-agency bargaining processes in shaping government decisions. In the I/M policy-setting process, however, the central actors were career staff at OMS with a long history of monitoring I/M program implementation. Indeed, working under tight, court-ordered deadlines, and hard-pressed to meet an array of rulemaking mandates, EPA offices throughout the agency were struggling to meet their own CAAA deadlines, let alone provide each other with comprehensive assistance on complicated rulemaking efforts.

## Cohen, March, and Olsen's Garbage Can Model of Organizational Choice and Kingdon's Policy Windows of Opportunity

In 1972, Michael D. Cohen, James G. March, and Johan P. Olsen published a seminal discussion of large bureaucracies titled *A Garbage Can Model of*

*Organizational Choice.* The Garbage Can Model postulated that policy setting occurs at the intersection of problems, solutions, participants, and decision opportunities (Cohen et al. 1972).

John W. Kingdon revised and extended the Garbage Can Model with his 1984 publication of *Agendas, Alternatives, and Public Policies,* in which he argued that federal policymaking occurs at the confluence of three streams: problem recognition, policy generation, and politics. A "window of opportunity" opens when the streams intersect. "But," Kingdon noted, "the window is open for only a while, and then it closes." He argued that when policy windows open, solutions emerge from long-standing options debated and shaped over time; he called these "the short list of ideas" (Kingdon 1984, *94, 146–149*).

Kingdon also observed that elected officials and their appointees shape the policy agenda. However, he found that career staff largely defined the policymaking alternatives available. Perhaps one of the most important emphases of the Kingdon model is its focus on politics. Kingdon cautioned, "In the political stream, if one does not pay sufficient attention to coalition building through bargaining, one pays a major price" (1984, *168*).

Interestingly, the Kingdon model does not account for the rapid ascension of RSD's influence. At the time of the I/M debate, RSD was a recent invention, not a policy option "honed in the political stream." However, RSD was perceived as so potentially valuable that it quickly jumped to, as Kingdon might have described it, the short list of ideas being debated. RSD gained prominence not just because it was innovative, but because it leaped into a policymaking vacuum: the lack of scientific consensus over how to identify high-emitting cars.

The concepts espoused by Kingdon, and Cohen et al. before him, offer two important insights concerning Smog Check. First, the concept of a policy *window of opportunity* sheds light on EPA decisionmaking. Prior to the court-ordered 1992 deadline to issue I/M regulations, the agency's last opportunity to establish I/M policy had been in response to the 1977 CAAA. EPA career staff knew that the 1992 regulations represented a rare window in which they could effect change.

The second insight concerns politics. EPA developed its I/M regulations through notice-and-comment rulemaking, then attempted to persuade stakeholders that its policy made sense. However, as Kingdon noted, "consensus building in the political arena, in contrast to consensus building among policy specialists, takes place through a bargaining process rather than by persuasion" (1984, *171*). The notice-and-comment process was more analogous to reaching consensus among policy specialists than to bargaining with legislators.

During the Smog Check debate, EPA staff, particularly at the request of the agency's political appointees, invested heavily in press outreach efforts that attempted to sway public opinion and persuade elected officials to accept EPA's

I/M policy. The next chapter examines the efficacy of those efforts and probes further into the roles of the public and the press in environmental policy setting.

## NOTES

[1] Notwithstanding the time constraints imposed on EPA, some in the air quality community argued for more research. For example, researcher Douglas Lawson, one of the first investigators to employ RSD units as a research tool, advocated as early as 1993 for an extensive pilot program to better inform the I/M debate. Lawson called for an I/M "shootout" to test "as many as nine different I/M program options, plus one control group" in different parts of the country for at least three years (Lawson 1993, *1574*). One can make a case that the time constraints imposed by the 1990 CAAA were not a root cause of the I/M conflict. EPA was not hesitant to acknowledge openly when statutory deadlines were impossible to achieve, and to allow alternative deadlines in contradiction of statute. For example, in the 1992 I/M regulations, EPA noted the impossibility of meeting the CAAA's deadlines and instead gave states several more years for implementation (U.S. EPA 1992e). Had EPA successfully completed scientific research in advance of the 1990 CAAA, the agency arguably might have been able to build technical consensus for an I/M policy, even within (or perhaps in spite of) the compressed schedule mandated by Congress. Chapter 11 includes additional perspective about the importance of the 1990 CAAA and their role in the Smog Check debate.

[2] The I/M rulemaking process was similar to what the agency employed when implementing regulations involving corporations or air quality management professionals. An example from the same time period as the Smog Check conflict involves EPA's 1993 promulgation of the transportation conformity regulations. The 1990 CAAA mandated that regional transportation plans conform to the same assumptions and goals as the air quality plans adopted for that same metropolitan area. EPA detailed these "conformity" requirements in a federal regulation that was implemented by transportation and air quality planning professionals in federal, state, and regional government agencies, as opposed to the general public or elected officials.

[3] Countries can also have tendencies when addressing environmental issues. In contrast to the U.S. command-and-control style, one researcher noted, for example, that Great Britain's approach was "flexible, informal, consensual, and incremental," and driven more by scientific proof and the need for demonstrated cost-effectiveness. Germany, on the other hand, regulated environmental problems by taking more of a precautionary approach (sometimes referred to as the *precautionary principle*), employing best available technology to address potential environmental problems (Wurzel 2002).

CHAPTER 9

# THE PUBLIC, THE PRESS, AND ENVIRONMENTAL POLICY

Among the many issues raised by the Smog Check story are questions involving the roles of the public and the press in setting and implementing environmental policy. In 1992, under pressure from a court-ordered deadline to issue enhanced I/M regulations, EPA used standard rulemaking approaches to solicit public feedback on its proposed I/M rule, and then issued its final regulations. The process, hurried as it was, allowed little opportunity for EPA to engage the public to measure the acceptability and practicality of its proposals. Yet as the agency noted when developing its enhanced I/M policy, "I/M programs need to be accepted and supported by the public to be successful" (U.S. EPA 1992e). Although the agency met its court-ordered schedule, the resulting Smog Check conflict, and later real-world implementation problems, substantially altered and delayed programs around the United States.

In contrast to the limited time EPA engaged the public while formulating its I/M rule, it worked extensively to engage the press while the Smog Check conflict took place. The agency's press efforts sought to shape news coverage and win support for EPA's policy among California's elected officials. Although this press outreach influenced media coverage during the Smog Check debate, the efforts failed to produce the agency's desired policy outcome.

Drawing on insights from the literature and the Smog Check conflict, this chapter offers some brief observations about the efficacy of public and press involvement in technically complex environmental policy setting. As discussed here, an ongoing need exists to engage the public in environmental policy setting and implementation. Although initially costly, public participation, via elected representatives, individual members of the public, or other representative groups or organizations, has the potential to smooth policy adoption, reduce implementation delays, and improve policy outcomes. In addition, proactively working with the press has important benefits, especially with regard to educating the public and the elected officials who represent them; these benefits, however, do not necessarily translate to influencing policy outcomes.

## THE ROLE OF THE PUBLIC

The literature shows that public participation offers unique opportunities to improve the viability and quality of policy outcomes. The U.S. National Research Council (NRC) has found that the process of developing policy is advanced when the participants represent a broad spectrum of stakeholders. According to the NRC, "science advances by deliberation and not just by analysis," and that deliberation, through peer review or other processes, is essential to "uncover errors and deepen understanding" of the policy problem and its potential solutions. "Involving the spectrum of interested and affected parties in deliberation can make the process ... more democratic, legitimate, and informative for decision participants" (NRC Committee on Risk Characterization 1996). The NRC added that deliberation could also improve the acceptability of policy decisions, a finding observed by others as well. For example, in a finding directly reminiscent of the Smog Check experience, the European Commission (EC) noted that "consultation and democratic scrutiny require time, and this might conflict with the need for quick decisions." On the other hand, the EC found "experience shows that quick decisions reached without scrutiny and consultation might prove socially unacceptable and thus inefficient" (Liberatore 2001, *7*). Striking a similar theme, policy researchers Renée Irvin and John Stansbury noted, "When the political situation is volatile and top-down decision making would be unpopular (if not unworkable), the up-front cost of citizen participation may be worth the additional funding because the costs of a difficult implementation of policy might be avoided" (2004, *58*). Public policy professor Nancy Roberts also cautioned that *not* including the public in policy setting risks "the potential for implementation disruptions and failures, costly litigation when citizens challenge administrative actions, not to mention lost good will and opportunities for social learning" (2004, *339*).

A consistent theme in the literature is the integral link between democratic governance and public participation. The EC noted, for example, that "democracy depends on people being able to take part in public debate" (Commission of the European Communities 2001, *11*). Irvin and Stansbury identified various benefits that accrue to both government and citizens from greater citizen involvement, including improved policy outcomes, avoided litigation, and the ability to break policy deadlocks. They noted that government administrators, "through regular contact with citizens who might otherwise not be engaged in the policy process, learn which policies are likely to be explosively unpopular and how to avoid such policy failures" (2004, *56*).

## Overcoming Hurdles to Public Involvement in Environmental Decisionmaking

Despite the culture of public involvement in democratically organized societies, the technical complexity of environmental problems has discouraged some decisionmakers from fostering public involvement. The NRC observed that agencies often perceive that complex issues need to be dealt with solely by experts (NRC Committee on Risk Characterization 1996). Roberts canvassed the public participation literature and identified dilemmas that must be addressed for participation to work successfully. One dilemma involved the handling of technically complex issues, which Roberts framed as the question, "How can ordinary citizens participate in the decisions made about complex technologies, especially when there can be wide disagreements among the experts . . . ?" (2004, *326–327*).[1]

The literature shows, however, that public participation is necessary even in the context of complex decisionmaking. Researcher Richard Sclove noted that when evaluating the merits of various policy options for a complex problem, ordinary citizens are in a far better position to protect their freedom and democracy than are elite experts, who might be focused more on the technological merits of a particular solution (1995). A more recent study of the community engagement literature found that "broader civic participation in deliberative processes to consider important social issues is recommended for addressing deep and complex problems" (Head 2007). The NRC pointed out that by excluding public participation, agencies risk alienating their constituents and encouraging their interference with the policy outcome. They argued for agencies to seek public participation across the spectrum of stakeholders, at every step of the process: "The common practice of eliciting comments only after most of the work of reaching a decision has been done is a cause for resentment" (NRC Committee on Risk Characterization 1996, *78*).

The use of pilot projects offers an opportunity to test programs for public acceptability and practicality. As some have noted, the social consequences of complex decisions are hard to predict, and pilot projects can produce real-world insights to inform policy. Sclove argued, for example, that only through pilot projects, or "voluntary social experiments," as he termed them, will it be possible to discover social consequences that are otherwise difficult to anticipate (1995, *55*).

In the years following the Smog Check conflict, opportunities to engage the public increased tremendously as access to the Internet became widespread. EPA received some 300 comments on its enhanced I/M regulations in 1992. Four years later, when EPA proposed to update the National Ambient Air Quality Standards for PM, the agency, aided by the Internet, fielded more than 50,000 public comments. In the Internet era, the challenge for public officials has become more an issue of how, rather than whether, to facilitate public input.

Some caveats are necessary, however. Despite the general case to promote public involvement, public participation is unnecessary, impractical, or ineffective in certain situations. Irvin and Stansbury (2004), for example, documented an unsuccessful effort to promote public participation in a land use and water management case. They found that a lack of any perceived crisis or urgent problem, combined with a public role that was merely advisory, demotivated public interest in the policy-setting process. The NRC also cautioned that seeking the deliberative input of stakeholders will not eliminate all the controversy associated with highly contentious decisions. The more that is at stake, and the more that "values and interests conflict," the less likely a decision will be widely accepted. Some parties will agree to deliberation simply to delay an outcome, because they benefit from the status quo (NRC Committee on Risk Characterization 1996).

## Observations Related to the California-EPA I/M Experience

The Smog Check policy experience involved a progression of steps to engage the public. During the 1992 rulemaking process on its national I/M regulations, EPA received and responded to public comments, mainly from air quality professionals. As the California-EPA debate unfolded, a second level of public participation took place: interactions with elected officials. Also, as California implemented its enhanced I/M program, Smog Check directly engaged the public through vehicle inspections. As noted in earlier chapters, California's initial implementation resulted in long inspection wait times, angry motorists, state legislative action to modify the program, and delayed implementation. Finally, California state senator Newton Russell provided a long-term platform for public participation in the Smog Check program when, in 1993, his legislation modified the California IMRC. Russell's changes, which altered the IMRC from a panel of air quality professionals to one that included representatives from various stakeholder groups, expanded Smog Check oversight beyond the technical experts who viewed the program through a narrow lens.

There is a striking contrast between the literature's arguments for early public participation and the incremental progression toward increased public involvement in the Smog Check experience. EPA's prescriptive 1992 policy retained virtually all I/M policy-setting authority for the federal government, rather than delegating responsibility to lower government levels; this limited opportunity for public involvement at the state and regional levels. However, public input increased once EPA engaged in political bargaining with California's elected officials.

It is understandable that EPA officials believed they lacked the time for an extensive public participation process in setting I/M policy, given the agency's court-ordered mandate. However, as supported by the literature, time spent seeking public input during policy setting can ultimately save time during implementation, a relevant point given the I/M debate outcomes.

The difficulty of anticipating real-world outcomes suggests that California and EPA would have benefited from additional pilot projects—limited implementation in selected metropolitan areas, for example—to solicit public response to the test-only provisions.[2] As others have noted, policies that rely on public participation become more complicated and politicized, and their implementation takes longer (Goggin et al. 1990). Pilot efforts, Sclove's social experiments, might have identified implementation bottlenecks. As public policy professor Eugene Bardach observed, "Policy ideas that sound great in theory often fail under conditions of actual field implementation. The implementation process has a life of its own" (1996, *32*).

## THE ROLE OF THE PRESS

During the Smog Check conflict, rather than engage the public to help design and implement policy, EPA sought to persuade the public, via their elected representatives, to adopt and implement the agency's model I/M program. A key component of EPA strategy was to communicate agency positions via press outreach. EPA's press efforts during Smog Check were certainly not unique among government agency actions. As others have pointed out, for example, "the White

House invests substantial energy and time in attempting to shape the media's attention" (Edwards and Wood 1999, *328*). Anecdotes can illustrate the potential power of media coverage on policy outcomes. For example, Powell (1999) recounted the 1983 discovery of dioxin in fish living downstream of U.S. pulp and paper mills. The discovery quickly led to placing dioxin on EPA's research agenda; however, it was not until the 1987 publication of a *New York Times* article on the dioxin problem that the issue was transferred to the agency's regulatory agenda. Given the potential importance of media coverage, it is not surprising that EPA focused energy on media relations during the Smog Check debate, though, as described here, the impact of those efforts was limited.

Many of EPA's Smog Check press actions were motivated by the premise that the agency could shape news coverage and thereby influence state legislators directly, or at least indirectly via public opinion. This premise is investigated in the following discussion, which begins by highlighting the sometimes tenuous relationships found in the literature among the media, public opinion, and policy outcomes, and then focuses on press actions during Smog Check.

## Media Influence on Public Opinion

In his seminal work on what shapes policy, University of Michigan professor John Kingdon noted that the "mass media clearly do affect the public opinion agenda," and that the public's attention to issues tracks media coverage (1984, *61*). Another analyst observed that when an issue is highly technical, and the general public lacks an understanding of the policy questions being debated, the media can play a strong role in shaping the public's understanding and perception of the issue (Nelkin 1987).

Political scientist Donald Jordan investigated the effect of *New York Times* news stories, focusing predominantly on front-page articles. He found that within one to seven months after publication, individual articles reporting news and expert views changed public opinion by approximately 1 to 2 percentage points—findings comparable to those of prior work that documented even larger impacts from television news. "Many news stories, each having a small effect, could add up to a substantial impact," said Jordan (1993, *197*). A later analysis of the literature found that public opinion about issue importance correlated with the amount of news coverage an issue received, that people think problems are more important if the media they listen to or read most devotes more time to those problems, that increased news coverage of a problem precedes measured increases in the importance of that problem to the public, and that people exposed to news stories about an issue are more likely to find that issue to be important. The same analysis concluded that media coverage shifts public opinion when it alters beliefs, confers new knowledge, or changes perceptions of societal consequences (Krosnick et al. 2006).

## Media Influence on Policy

Despite the close association between media coverage and public opinion, the connection between news coverage and policy outcomes is less clear-cut. Kingdon's work in the 1970s and 1980s on health and transportation issues found that media impacts on policy agendas could be found, but they were rare. Kingdon said that both policy practitioners and journalists agreed that media coverage is ephemeral and lacks the staying power to influence most issues. Despite periods of intense press attention to a particular issue, many policy practitioners wait out the media interest and pursue their policymaking efforts to completion after the press has moved on to cover other crises. Despite these limitations, Kingdon observed that the media have an information-sharing role during policy debates. For example, Kingdon noted that the media act as a communicator within the policy community: a story that makes it into the newspaper or onto television can gain the attention of senior policymakers when routine memoranda and reports fail to catch their interest (Kingdon 1984).

Later work by political scientists George Edwards and B. Dan Wood found that, "anecdotal evidence aside, we know very little about the influence of the media on the policy attention of public officials." In response, Edwards and Wood examined five policymaking situations from the Reagan and Clinton U.S. presidencies: crime, education, health care, U.S.-Soviet relations, and the Arab-Israeli conflict. They assessed relationships among television coverage, congressional actions, and presidential actions and found mixed results: the presidents responded to and caused media attention on educational issues, responded to media coverage on foreign policy issues, and shaped but did not respond to media coverage of health care issues; the U.S. Congress was unaffected by, nor did it affect, media coverage in any of the five cases (Edwards and Wood 1999).

## Assessing the Link: Influence of Public Opinion on Policy

The record is mixed on whether public opinion affects policy outcomes. Researchers from Northwestern University assessed the literature on the link between public opinion and policy and observed that only under some conditions and for some issues is the link strong. They noted, for example, that some issues were more "opaque" to the public, giving policymakers more room to maneuver; responsiveness to public opinion also varied by whether policies were addressed by elected officials or by others who were shielded from a loss of office (Manza and Cook 2002).

Sociologist Paul Burstein later found that the overall influence of public opinion is limited. In a random sampling of 60 policy proposals before the U.S. House of Representatives, public opinion data were available for only 36; of these, policy actions were consistent with public opinion about half the time (Burstein 2006). Interestingly, Burstein's findings were similar to the earlier observations of Kingdon, who determined that public opinion was "neither insignificant by any means, nor among the most prominent in the total array of sources, but just about in the middle" (1984, 69).

Burstein noted that in 14 of the 60 policy proposals he investigated, no public preference could be found. He therefore concluded that the relationship between public opinion and policy actions is overstated in the literature because of sampling bias—most studies focus on higher-profile issues for which the public has an opinion. Based on the random sample of issues he evaluated, Burstein found that government actions matched expressed public opinion in only 18 out of 60 cases (2006). Burstein's work is consistent with observations made by others. Manza and Cook, evaluating U.S. congressional efforts, said that despite the "startling increase in polling" data that has become available, opinion data do not exist on most questions that policymakers address. Of the opinion data that is available, much of it cannot be readily accessed by elected officials. The authors credited the media with helping frame public opinion by deciding what small fraction of available polling data to report (Manza and Cook 2002).

In summary, the media appear to have more influence on public opinion than on policy outcomes. Public opinion, when expressed, correlates with policy actions only some of the time. The implication is that there are important limits to how much media coverage can shape policy outcomes by shifting the opinions of legislators or the general public.

## EPA, the Media, and Smog Check

During the California-EPA I/M debate, EPA pursued a mix of goals regarding the press. First, the agency acted under the assumption that the press shaped the opinions of the public and state legislators. Thus it was important to have the press write stories favorable to EPA's position. Second, the agency wanted to ensure that the press was knowledgeable about and could convey EPA's positions, to avoid mischaracterizations of agency policy. Third, the agency wanted to shape the perceptions of those outside of California. Policymakers from other states were watching the California conflict, and although EPA explained to these officials that the agency's support of its I/M policy remained steadfast, officials outside the state were concerned that EPA would relax its I/M policy to give California special latitude. Thus it helped EPA to have press coverage about California that confirmed the agency's policy stance.

EPA attempted several tactics to influence press coverage. First, it mounted outreach efforts to educate the press with briefings for journalists and editorial boards. The agency also routinely developed talking points in advance of expected press contacts, to ensure strategic press communications and emphasize its position in the I/M debate. EPA also conducted preannouncement press briefings, typically the day before a major announcement. In exchange for these briefings, the press consented to delay publication of news stories until after EPA's official announcement. Agency staff also wrote articles for publication in major newspapers to communicate its message directly via the media. Finally, when EPA believed it was essential to try to shape press coverage for a particular event, the agency gave individual reporters advance access to the story, hoping that with an exclusive interview, the reporters would provide EPA with more favorable treatment in resulting news articles.

Obvious questions emerge: Was EPA effective at communicating the agency's position? Did its press efforts influence the policy outcome? Three conclusions emerge from the Smog Check experience. First, there is little doubt that EPA efforts to influence press coverage had an impact on media reporting. The media ran articles as a direct result of EPA actions, and when the agency briefed journalists and disseminated talking points, these articles usually included some rendition of EPA's version of the story.

A dramatic example involved the press coverage following EPA's abrupt decision to back away from the sanctions process on November 23, 1993. After deliberating until late evening about how to portray its decision to abandon sanctions, EPA decided to announce that significant developments in the I/M debate had occurred, necessitating a halt to sanctions. The agency issued a three-paragraph press statement with an explanation that "state leaders articulated a new willingness to work with us to forge a solution" (Nichols 1993). Following the announcement, the press reported: "The Environmental Protection Agency said that the potential breakthrough was prompted by what it called a 'new willingness' on the part of state legislators to end the months-long standoff. State officials, however maintained ... that Gov. Pete Wilson's aides hadn't signaled a new position" (*Wall Street Journal* 1993). As illustrated by this excerpt, the press conveyed EPA's position but also presented alternative views. Therefore, although EPA had some success in steering the content of the story, the agency was limited in its impact on the overall press discussion presented.

Another example illustrates the consequences that resulted when EPA did not attempt to shape press coverage. In March 1994, once EPA and California negotiated an agreement, the agency did not proactively communicate the agreement to the press. The resulting press coverage gave erroneous accounts of the California-EPA deal. Both the *San Francisco Chronicle* and *Los Angeles Times* incorrectly reported, for example, that the agreement was a commitment to send only 15 percent of vehicles to test-only stations, a point true only for the first year of program implementation.

Second, press accounts no doubt did have an impact on state legislators and other government officials involved in the I/M policy debate. For example, early in the California-EPA debate, State Senator Robert Presley encouraged EPA (via direct communication with the author) to continue generating press coverage saying that the agency was taking a firm position regarding Smog Check. Presley believed the press coverage helped demonstrate to his fellow legislators that EPA was serious about its intention to replace the existing Smog Check program with something better.

Third, EPA's influence changed over the course of the debate. Initially, reporters knew little about the technical strengths and weaknesses of the various I/M options being evaluated. In the earliest stages of the I/M issue, journalists were more inclined to report information as conveyed by EPA, before they had had an opportunity to investigate deeply the merits of the agency's position. However, as the debate developed, and particularly as the high stakes of the outcome became more apparent, the media collectively began to conduct more in-depth investigative and analytical reporting. In effect, journalists went through an issue-learning process the longer the debate lasted. The important implication is that once they understood the issue, investigative reporters could more readily discern between agency spin and fact. Nine months into the California-EPA conflict, for example, an article by Scott Thurm that penetrated to the technical shortcomings of EPA's policy appeared in the *San Jose Mercury News* (1993a). By March 1994, Thurm had written at least nine articles on Smog Check. Once journalists like Thurm had digested the broader sweep of opinions about Smog Check, it became much more difficult for EPA to influence media coverage to favorably portray the agency's policy positions.

An example illustrates EPA's limited ability to influence the press once journalistic learning had taken place. On August 26, 1993, EPA administrator Carol Browner granted an exclusive interview to Melissa Healy of the *Los Angeles Times*, in the hope of influencing the reporting of the agency's decision to abandon its rigid policy stance in the wake of the RAND Corporation findings. Following Browner's interview, the *Times* ran a lengthy Smog Check article. However, by that point in the debate, the paper had already published several articles on Smog Check. Healy's article, published in collaboration with another *Times* reporter, presented various stakeholder viewpoints, indicated that many believed EPA had compromised as a result of political pressure, and referenced the RAND study as finding little technical support for the agency's policy. On the other hand, the article did present an in-depth description of EPA's version of the I/M story. Thus while not diminishing the newspaper's portrayal of the perspectives held by various Smog Check stakeholders, the exclusive interview helped ensure that EPA's story was presented along with those of other debate participants.

In summary, an important Smog Check insight is the need to educate the press to help shape the content of news stories. Equally important, though, is the insight that press outreach had only a limited impact on reporting and policy-setting. EPA attempts to shape news coverage yielded reduced benefits once journalists were well informed about the story. Notably, as the debate neared its resolution, EPA's many press efforts did not yield the desired reward of a policy outcome that matched agency goals. Despite these limitations, EPA political appointees at the time believed that press outreach should have contributed more than it did to persuading the general public, and California's leadership, to accept the agency's I/M policy.

The challenges of working with the press and the public were just two of the many difficulties encountered during the Smog Check conflict. As recalled by former EPA regional administrator Felicia Marcus many years later, Smog Check was a "perfect storm" of regulatory problems (see Chapter 11). The next chapter extracts lessons from this storm and offers insights to set better regulatory courses in the future.

## NOTES

[1] Other dilemmas cited by Roberts involve dealing with complex risks; overcoming the large scale or size of the "administrative state," including excluded or oppressed groups; dealing expeditiously with time constraints and crises; and contributing to the common good through deliberative assessment (2004, *326–327*). Researcher Brian Head (2007) distilled from the literature six "learning challenges" to those participating in collective processes, including the need to build relationships and trust, develop and refine common directions and objectives, give up some demands for control, facilitate rather than direct, deal with "reform fatigue," and address the "two hats" problem of reconciling the dual roles of serving among collaboration participants and within a home organization.

[2] Former EPA official Dick Wilson noted in Chapter 11, however, that during Smog Check, the 1990 CAAA deadlines precluded the possibility of pilot projects.

CHAPTER 10

# LESSONS LEARNED AND A NEW REGULATORY TOOL

In 1984, the U.S. Supreme Court ruled on a Clean Air Act case in what has since become one of the most cited U.S. administrative law judgments: *Chevron U.S.A., Inc. v. Natural Resources Defense Council, Inc.*[1] In *Chevron*, the Supreme Court stated that reasonable actions by the U.S. Environmental Protection Agency, and by extension other government agencies, are entitled to deference: "When a challenge to an agency construction of a statutory provision, fairly conceptualized, really centers on the wisdom of the agency's policy, rather than whether it is a reasonable choice within a gap left open by Congress, the challenge must fail. In such a case, federal judges—who have no constituency—have a duty to respect legitimate policy choices made by those who do" (467 U.S. 837, 1984).

In other words, EPA has the discretion to decide what is reasonable when interpreting its many statutory mandates, including those like I/M, where Congress tasks the agency to provide guidance to states. One might imagine that given such legal latitude, EPA would have tremendous leverage to ensure the successful advance of its policies. Yet, as illustrated by Smog Check, legal authority alone is not sufficient to secure policy implementation—administrators need to weigh other factors to avoid the problems presented here.

While many lessons can be learned from the U.S. I/M experience, the heart of this case study boils down to two fundamental observations. First, in the technically complex arena of regulating environmental problems, agencies have to do their scientific homework. Although the use of scientific information has important limitations (discussed later in this chapter), agencies must try to build a consensus within the technical community concerning the scientific understanding of the problem to be addressed and the likely benefits of the control options available. Control policies, once implemented, must enable the ongoing collection and assessment of scientific information to foster improvements over time—particularly when consensus is difficult to reach at the outset.

Second, the applicability of traditional command-and-control federal regulations also has important limitations—especially political ones. The command-and-control regulatory model doubtless has produced remarkable achievements over time. To cite just a few air quality examples, by virtue of federal mandate, automakers manufactured vehicles that consumed less fuel and emitted fewer pollutants, oil companies and electric utilities built new industrial facilities with state-of-the-art emission controls, and chemical companies phased out the production of harmful compounds such as ozone-depleting substances.

But the command-and-control model does not work in all situations. The more localized the implementation of a control becomes—for example, the more it involves controlling individuals rather than nationally distributed products—the more difficult it becomes to require its imposition via federal mandate. The Smog Check debate's political friction reflected the limitations of the command-and-control paradigm. As witnessed by Smog Check, some regulatory situations inherently call for greater flexibility. Misinterpreting these situations triggers risks—such as damaging political confrontations—that can delay progress toward environmental goals, a situation exemplified by Smog Check and aptly characterized by the political artwork of the debate period (see Figure 10-1).

Hindsight makes clear where, during Smog Check and its aftermath, EPA ventured in the wrong direction. It is prudent to ask, what might EPA do differently in the future? This chapter examines some of the progress made by EPA to avoid the problems exemplified by Smog Check and its outcomes, and then presents ideas to help structure future regulations that are both technically and politically acceptable.

## THE CHALLENGE OF AVOIDING SIMILAR PROBLEMS

Before this work presents its final recommendations, it is essential to reflect on the fact that problems like Smog Check emerge because the root causes are difficult to

**Figure 10-1. Cartoon illustrating the Smog Check dispute, from the *Sacramento Bee*. (© J.D. Crowe. Reprinted with permission)**

address. In the years following the California-EPA I/M debate, the agency made several efforts to address the core problems discussed in this study. However, a review of EPA experience with two key issues—government reinvention and use of scientific information—helps temper expectations about resolving some of these policy challenges.

## Government Reinvention

Largely as a result of several factors, including the 1993 Clinton–Gore National Partnership for Reinventing Government, the March 1994 California-EPA I/M deal, the November 1994 congressional election that gave Republicans control of the U.S. House and Senate, the December 1994 meeting between EPA administrator Browner and state governors upset over the I/M policy, and the March 1995 congressional hearings on I/M policy, EPA launched a comprehensive effort to reinvent how it conducted its business. The agency identified 25 high-priority reinvention actions, among them the goal to "encourage regulatory negotiation and consensus-based rulemaking." By February 1997, EPA had committed to create an Office of Reinvention. The new office's primary focus was to change the agency's culture so that staff would become more open to innovative approaches and rely less on the command-and-control regulatory tradition.

Two years into the reinvention effort, the U.S. General Accounting Office (GAO) assessed EPA's progress. GAO found that "while EPA has made some progress ... the agency still has a long way to go in resolving several key issues if environmental regulation is to be truly 'reinvented.'" Some of the main challenges GAO found were that reinvention among the agency's rank-and-file career staff would be difficult and take time, and that the statutory frameworks from which EPA drew its authority had influenced many of the regulatory and behavioral practices the agency sought to reinvent (U.S. GAO 1997).

By 2001, GAO had performed another review, assessing whether EPA was meeting its management challenges. One of these challenges was to shift the focus of decisionmaking from activities and processes to results and outcomes. A central management challenge for EPA, as pointed out by GAO, was for the agency to improve its working relationships with the states. GAO found that "EPA's relationship with its state partners has been characterized by fundamental disagreements over roles, priorities, and the extent of federal oversight." In its 2001 review, GAO found that EPA had made progress in repairing its relationships with states but needed to continue to improve (U.S. GAO 2001).

In January 2004, EPA released an annual report with findings by the EPA Office of Inspector General (OIG). The report noted that the agency had spent considerable effort attempting to reinvent its regulatory approach and acknowledged that GAO had previously cited reinvention as a management challenge for EPA. OIG cautioned, however, that statutory obstacles would have to be overcome for reinvention to succeed. It recognized that EPA had made several efforts to improve its working partnerships with states, but it also pointed out that EPA's relationships with states remained strained, and that disagreements continued regarding oversight responsibilities; priorities and budgets; and results-oriented performance measures, milestones, and data (U.S. EPA 2004b).

The record indicates that, much as GAO predicted in 1997, EPA reinvention required a long-term effort, and progress was limited in the first decade or so. As one researcher noted, as of the early 2000s, many states had been disappointed in the results of EPA's reinvention efforts: "It is still difficult to find, in many instances after more than a decade's experience with some initiatives, clear consensus among the major stakeholders regarding the success of any showpiece reform. Indeed, it may be difficult ever to achieve such a clear verdict" (Rosenbaum 2003, *194*). A separate study by policy analyst Barry Rabe found similar results, but in addition to citing EPA's difficulties in reinventing itself, it also cast blame on the states for their reluctance to be held to stringent performance goals (Rabe 2003). Others have noted that reform efforts begun under President Clinton failed to survive the transition to the next administration. One academic research team stated that "the reinvention movement was largely doomed with the election of President George W. Bush" and pointed out that Bush replaced the Clinton-era reinvention effort with his own reform program, the President's Management Agenda (Bertelli and Lynn 2006, *157*).

## Improved Use of Science

EPA's 2003–2008 strategic plan began its science discussion by stating, "EPA has identified reliance on sound science and credible data among the guiding principles we will follow to fulfill our mission to protect human health and the environment" (U.S. EPA 2003c, *153*). The agency took numerous steps following the Smog Check debate to improve its use of science. During its March 1995 government reinvention efforts, for example, one of its high-priority goals was establishing an EPA center for environmental information and statistics to improve access to the scientific information needed to support policy. The agency also identified other significant reinvention actions, including "setting priorities based on sound science" and commissioning an independent study to improve EPA's collection and use of information (U.S. GAO 1997).

Yet, as discussed by Powell (1999), as of the late 1990s, numerous factors impeded EPA's use of science during regulatory decisionmaking processes. Responding to requests by Congress and EPA itself for an independent assessment of the agency's use of science, the U.S. National Academy of Sciences (NAS) published a study titled *Strengthening Science at the U.S. Environmental Protection Agency.* The NAS observed that EPA had made progress in its use of science, especially since 1992 (the year the agency released its controversial I/M regulations). However, the NAS concluded that a large number of problems remained, as documented by Powell and others, and provided the agency with 15 recommendations, including creating a position of deputy administrator for science and technology, continuing multiyear planning for research, and improving the agency's peer-review policies (NAS 2000).[2]

In late 2000, a bipartisan group of 33 representatives from academia, environmental think tanks, state and federal government agencies, elected officials, and industry participated in a workshop to identify needed EPA management improvements and recommend an action plan to President-elect George W. Bush. The bipartisan group identified 11 major management challenges facing EPA, including "managing the development and analysis of environmental science, as well as using science to inform its decision making." The group's recommendations included improving the credibility of EPA's scientific work and reevaluating EPA regulatory and peer-review processes (DeMaio 2001). Some of the participants believed that politics was such a dominant factor in EPA decisionmaking that the science problems the agency experienced would never be resolved.

Powell, the NAS, and the bipartisan working group have not been alone in their criticism of EPA's use of science and data. In a 2001–2003 assessment, EPA's OIG identified "improving the quality of data" used by the agency as one of its key management challenges (U.S. EPA 2004b). Former EPA general counsel E. Donald Elliott also argued that "science is underrepresented in policymaking at EPA" and advocated creating a chief science officer position at EPA, empowering agency scientists to make policy recommendations, and including disinterested nongovernmental scientists in the administrative process (Elliott 2003).

Reflecting on how pivotal the lack of sound science was to the Smog Check conflict and EPA's management of I/M in later years, and how so many have observed EPA's scientific shortcomings for so long, it is reasonable to hope that the agency will improve its use of science in the future. Yet expectations must be tempered—there are limits to what science can resolve. Scientific information is usually a far cry from absolute truth. To use Sheila Jasanoff's words, scientific information is "socially constructed," meaning scientists bring their own biases to the work they do and the findings they present. Jasanoff identified how scientific information can be subject to manipulation and interpretation at all points in the research and policy analysis stages. She cited, for example, the limitations of peer review as a tool to ensure scientific integrity, as well as the role negotiations play in shaping science advisory board findings (Jasanoff 1992). Trevor Pinch, an academician, cautioned that science has a human side to it: although scientists are experts, they reach their expert conclusions in a "messy way" that may not be a transparent reading of the data available. Pinch noted that scientific findings include a great deal of uncertainty and likened scientists to professionals practicing their craft; they do their best but sometimes get it wrong (Pinch 2000). Judith

Layzer as well cautioned that in most environmental disputes, the science is highly uncertain (Layzer 2002).

Pinch and Jasanoff both discussed the concept (first named by Harry Collins) of Experimenter's Regress, where experiments rarely are duplicated exactly, and thus findings are difficult to replicate. Experimenter's Regress illustrates how scientists can arrive at diverging conclusions, leading one scientist or group of scientists to attack the findings of others. The result, as Jasanoff noted, is that, especially in adversarial processes such as formal rulemaking, scientific data are "deconstructed" in such a way that no experiment or finding stands up to inspection (Jasanoff 1990; Pinch 2000). A more caustic view was provided by Tulane law professor Oliver Houck, who cautioned against what he called the lure of good science: "Every lawyer knows what 'good science' is: the science that supports his or her case. All of the other science is bad" (Houck 2003, *1928*).

However definitive or uncertain the science, there remains a limit to its influence on the policy process. Resolution of regulatory disputes is based largely on legal, political, and economic criteria; science is not the main decisionmaking component (Powell 1999). Jasanoff articulated the concept well: "When the stakes are high enough, no committee of experts, however credentialed, can muster enough authority to end the dispute on scientific grounds" (1990, *234*).

What, then, to make of science and its role in the environmental decisionmaking process? Three limitations are apparent. First, because of budget and time constraints, EPA is limited in its ability to procure and use scientific information. Second, scientific information, once procured, is limited in its ability to provide definitive answers. Third, even if scientific data provide definitive insight, other decisionmaking factors may govern a policy outcome. EPA must continue to improve its use of science; as exemplified in this study, there are perils involved with using poor data. Policymakers, however, also must anticipate and prepare for an inevitable lack of definitive data. Thus the discussion must circle back to a concept introduced earlier: given the inherent uncertainty of environmental information, the technical consensus-building process will itself be a negotiation, no less so than the more obvious political processes used to develop policymaking compromises.

As the government reinvention and use of science discussions illustrate, any recommendations that emerge from the Smog Check story need to be tempered by practical considerations.

## FEDERAL VS. STATE ENVIRONMENTAL ROLES: IDEAS FROM THE LITERATURE

On May 5, 1998, Vice President Al Gore spoke to U.S. federal agency senior executives about the need for continued government reinvention; during his speech, he employed U.S. general George S. Patton's famous and often-quoted remark: "Never tell people how to do things. Tell them what to do and they will surprise you with their ingenuity" (Gore 1998).

Assuming the call for some form of government reinvention continues (taking new names, if not new direction, as administrations change), states will place ongoing pressure on EPA to delegate authority. For example, some have noted how, following the Clinton-era Smog Check challenges discussed here, tension between the states and the federal government increased during the tenure of U.S. president George W. Bush. One researcher characterized the George W. Bush era as a period when the federal administration worked to "recentralize oversight of environmental policy" even though states "demonstrated that they are not prepared to take a back seat to the federal government" (Rabe 2007, *429*). Such ongoing friction between the federal and state governments highlights the need to better define federal versus state roles and provide guidance on federal regulatory design.

Looking back, Smog Check illustrates that some situations are better suited to flexible, rather than prescriptive, federal mandates. Looking forward, an important question is which types of situations merit greater flexibility. Many others have explored federal versus state policymaking and implementation. The last part of this discussion samples the literature to identify broad themes regarding federal policy design, and then uses those insights, as well as insights from Smog Check and its outcomes, to construct a policy tool that reflects the lessons learned.

## Environmental Policymaking Roles

U.S. environmental policy specialist Paul Portney observed many years ago that there was no innate reason that environmental policies had to be federal. He recognized that states competed against one another to attract businesses, and the competition provided incentives to diminish environmentally stringent regulatory efforts.[3] However, Portney also observed that some environmental problems are geographically distinct and fail to transcend state boundaries. Such geographically limited issues, he argued, could well be addressed by the states in which they occur. Portney suggested that the need for federal policy be constrained to addressing multistate problems, such as acid rain; providing uniformity to businesses with operations in multiple states or whose products, such as automobiles, are sold throughout the nation; and sponsoring research on health effects, control strategies, and other environmental management issues (Portney 1990).

Walter Rosenbaum observed that as states have assumed a more active role in environmental management, they have accrued expertise and grown to be more competent stewards, especially since the 1980s. However, he noted, EPA and Congress have been slow to recognize the growth in state expertise. He suggested several actions that Congress and EPA should take to improve the agency's ability to protect the environment: Congress needed to pass a new act to organize and consolidate the myriad authorizations empowering EPA, provide relief from the unrealistic deadlines contained in the many statutes that direct EPA's workload, and appropriate more money to support EPA actions; and EPA needed to obtain more support from its regional offices and from states, and continue to change its operating culture away from command-and-control (Rosenbaum 2003).

Policy analyst Barry Rabe also acknowledged that delegation to states is often the best approach toward environmental regulation. However, Rabe, like Portney, pointed out that many air pollution problems transcend state boundaries and require a broader perspective for effective control. He also cautioned that states have demonstrated a wide variability in their compliance with environmental regulations and their ability to foster innovative environmental management. For example, he cited a 2001 study by the Resource Renewal Institute that ranked state capacity and commitment to environmental protection. Out of a possible 100 points, state scores ranged from a low of 8 points (Alabama) to a high of 73 (Oregon); California scored 42 (Rabe 2003, *Table 2-1*).[4] The implication is that if left to their own devices, individual areas arrive at a wide range of endpoints, not all of which are environmentally protective.

Rabe identified an effective federal role as requiring information to be released to the public; providing funds to help states implement programs; supporting research and development; facilitating multistate interactions; addressing national or cross-state problems; and offering on-site state assistance through EPA regional offices, including promoting the sharing of data and ideas among federal, state, and local agencies responsible for environmental protection. He supported having EPA, whenever possible, establish performance measures rather than command-and-control requirements (Rabe 2003).

Legal scholars David Adelman and Kirsten Engel said that "the central challenge for environmental federalism is limiting federal authority to a level that is not overly destructive of policy diversity and innovation" (2008, *1832*). They argued for what they called "Adaptive Federalism," which they defined as a "dynamic system of overlapping federal-state jurisdiction" over environmental protection. Their framework included three principles to better define the federal role: establishing a presumption against the federal preemption of state and local initiatives; discouraging federal actions (or "ceilings") that prevented states from setting policies more stringent than those of the federal government; and tempering the uniformity of federal mandates by allowing selected jurisdictions, such as states with demonstrated track records of leadership, the opportunity to advance independent policies.

## Environmental Policy Implementation Roles

The policy implementation literature offers additional insight regarding federal-state relationships. Michael Hill and Peter Hupe noted that implementation behavior is a function of what takes place at street level, implying that federal policy implementation will reflect the actions of state and local actors. In consequence, they observed the importance of negotiation and bargaining among policy actors, an overall process they referred to as the "co-production of policy," and said that "the idea that policy processes are in general an interplay between various actors and not centrally governed by government is now broadly accepted" (Hill and Hupe 2002, *77, 136*).

As early as 1989, researchers identified that "one of the best-documented findings in implementation literature is the difficulty of obtaining coordinated action ... among the numerous semiautonomous agencies involved in most implementation efforts" (Mazmanian and Sabatier 1989, *27*). However, the literature shows that cooperation among multiple partners is facilitated by less coercive, more flexibly designed policies. For example, Robert Stoker, author of *Reluctant Partners*, an examination of federal policy implementation, argued that to enhance protection of individual liberty, the U.S. system of government diffused authority across federal, state, and local levels. He concluded that federal programs require, therefore, the active cooperation of state and local implementation partners. Stoker observed several limits to the effectiveness of federally coercive approaches to implement policy, and he noted that cooperation is encouraged by defining "standards of conduct"—in other words, performance goals—rather than overly precise mandates. Importantly, he observed that federal policy is a "blunt instrument" and that federal policymakers should abandon the hope that they can design policy with "surgical precision." When dealing with complex problems, policy "change and adjustment are inevitable," he said; to induce cooperation among different levels of government, policies must allow implementation flexibility to foster needed policy shifts over time (Stoker 1991, *127, 182–183*).

Other researchers maintained that in situations involving new problems and where data are limited, government actions should be less coercive; they also observed that in situations where power is dispersed (e.g., the U.S. political system), policy actors have to convince others of the soundness of their preferred solution—they cannot simply exercise raw political power to achieve implementation (Sabatier et al. 1993). In *The Powers to Lead*, an examination of effective leadership in the public and private sectors, Joseph Nye struck a similar theme when he distinguished between what he termed "hard" and "soft" power. Hard power, he explained, is based on inducements or threats, whereas the exercise

| Ambiguity \ Conflict | Low | High |
|---|---|---|
| Low | *Administrative implementation*<br>Outcomes are well understood and are a function of the available resources to make them happen. | *Political implementation*<br>Policy is opposed and outcomes depend on political power and resources of agency imposing requirement. Coercion and financial tools are used to influence outcomes. |
| High | *Experimental implementation*<br>Because of policy ambiguity and resulting site-to-site variations, policy outcomes are hard to predict. | *Symbolic implementation*<br>Although seemingly implausible, high-conflict, high-ambiguity situations could exist if policies seek to redistribute power or goods. Outcomes are unclear. |

**Figure 10-2. Matland's Ambiguity-Conflict Model of Policy Implementation**
*Source:* Adapted from Matland 1995

of soft power involves attracting others to one's ideas. In the context of governing, greater diffusion of power emphasizes soft power approaches. "Soft power is a staple of daily democratic politics," argued Nye (2008, *30*).

## Building Support for Policies by Easing Conflict

A basic premise that emerges from the literature is the commonsense notion that the more prescriptive the government mandate, the less political support it is likely to garner. Political scientist Richard E. Matland illustrated this relationship with his Ambiguity-Conflict Model of policy implementation. This model explains policy implementation as a function of both the level of conflict that exists over a policy and the degree to which policy requirements are clearly defined. Matland theorized that in situations of high conflict and low ambiguity, implementation outcomes are determined by the political power of the conflicting parties. If one party has sufficient power to impose its will on the other, there is less need for ambiguity and political bargaining (Matland 1995). Figure 10-2 includes a representation of Matland's model, adapted here to help interpret the Smog Check experience.

Figure 10-3 presents an analogous model for I/M, organized as a conflict matrix illustrating command-and-control versus performance measures. Conceptually, as shown in the figure, an inverse relationship existed between the degree of specificity

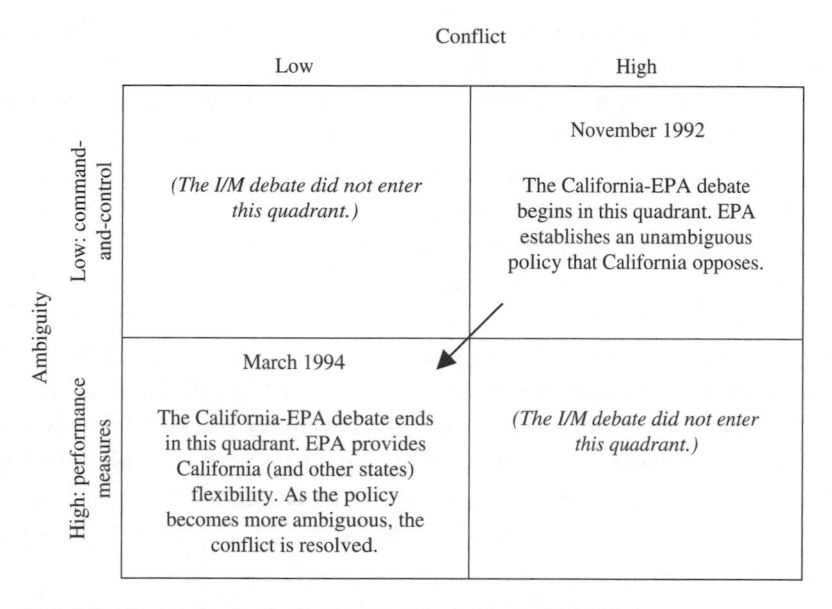


**Figure 10-3. A Modified Ambiguity-Conflict Model of the California-EPA I/M Debate**

(or lack of ambiguity) in EPA's I/M policy and the level of conflict the agency experienced when attempting to persuade California to accept and implement federal policy. Once EPA made its requirements more ambiguous, tensions eased and the conflict was resolved.

This command-and-control versus performance measures version of Matland's matrix is useful as a reminder that performance measures establish outcome goals but allow implementation flexibility (ambiguity), and thus they are more likely to foster collaboration over conflict. An important question is whether greater flexibility results in the desired outcome. The Ambiguity-Conflict matrix does not address the quality of resulting outcomes once tensions ease; it merely reinforces the commonsense understanding that conflict resolution is assisted by ambiguity. However, the matrix does address the variability in resulting outcomes. Matland described the low-conflict, high-ambiguity combination as experimental and likely to result in hard-to-predict outcomes that will vary from site to site. He also noted that ambiguity is not necessarily undesirable. Ambiguous policies can foster experimentation and continued learning within the policy community (Matland 1995).

## Summary Observations from the Literature

The literature on the appropriateness of federal versus state roles is extensive, and the foregoing discussion has provided but a brief synopsis. Nevertheless, key messages are apparent from this sampling of opinions. Some core environmental management activities logically fall to the federal government. Because state environmental management efforts continue to mature over time, the federal government should move away from its command-and-control tradition and toward a performance-based management approach when and where appropriate. Such a movement can ease federal versus state tensions and promote multiagency partnerships. The progression toward granting states flexibility, however, remains slowed by the statutory obligations imposed on EPA, the inadequacy of some state programs, lack of adequate funding, the nature of individual problems, and the agency's institutional reticence to shift from its traditional management role.

This study acknowledges that although systemic change would assist EPA in its efforts to protect the environment, it will likely be many years before such change is achieved. The recommendations cited by Rosenbaum, for example, argue for a wholesale change in the U.S. environmental statutory framework, as well as a need to continue to modify the deep-rooted cultural practices at EPA. Such changes will not happen quickly.

## JUDGING WHEN TO USE PERFORMANCE MEASURES RATHER THAN COMMAND-AND-CONTROL MANDATES

Given the difficulties illustrated by Smog Check and the broader U.S. I/M policy experience, new resources are needed to help managers structure regulatory mandates in light of the complicated federal-state relationship. In an assessment of Clean Air Act implementation a decade after the 1990 amendments, GAO observed, "One of the challenges facing the Congress in considering the Clean Air Act's reauthorization is determining the appropriate balance between traditional command and control approaches and more flexible approaches that allow state and local air pollution control agencies and other stakeholders to implement the most cost-effective strategies, while meeting national air quality goals" (U.S. GAO 2000, *17*).

This discussion contributes a new regulatory tool to help sort through the many factors that can lead policymakers to favor either performance-based or command-and-control directives. The new tool, introduced here and called E-REGS, for the Environmental Regulatory Spectrum, helps assess where appropriate management of an issue might fall on a spectrum that ranges from setting flexible performance-measure-based goals to mandating less flexible command-and-control-based approaches. The tool is not a cure-all for the systemic difficulties exhibited in the Smog Check story. As noted earlier, many factors contribute to a policy outcome; some factors, such as elections and economic conditions, are beyond the control of practitioners. E-REGS is a tool to help analysts weigh situations, understand what can and cannot be controlled, and gauge how to regulate.

Some take the view that virtually all government regulations are counter-productive. E-REGS offers little for those who protest against regulatory actions in general. It does, however, provide guidance for those who support the view that government regulation plays a necessary role in limiting environmental damage.

## E-REGS: The Spectrum Described

It is important to deliberate on a case-by-case basis whether performance measures or more prescriptive mandates are the appropriate management tool. For example, contrasted against the theoretical concepts supporting use of performance-based measures are the real-world implementation lessons associated with the Clean Air Act itself. In *Blue Skies, Green Politics*, an assessment of the passage and implementation of the 1990 CAAA, Gary Bryner observed: "In an ideal world, the EPA would defer to states' decisions about means, while ensuring that the goals were achieved. But the history of the Clean Air Act shows that states have often done little to remedy air pollution problems" (Bryner 1995, *225*).

Figure 10-4 shows how the E-REGS regulatory development tool structures the development of national environmental policy as a function of politics, problems, and policies. These three factors are derived from Kingdon (1984), who indicated that policies emerge when politics favor problem resolution, the problem is ripe for solving, and well-honed solutions become available. E-REGS makes explicit that the specific character of individual environmental problems, the political circumstances that exist at the time of the policy development process, and the attributes of specific policies all must be weighed in assessing the appropriateness of performance measures or other management approaches.

The E-REGS guide does not assign relative weights nor provide a quantitative approach to issue evaluation. Each environmental issue is unique, and at any point in time, one criterion (for example, political considerations) may dominate in importance over others. However, analysts can use E-REGS as a qualitative guide to organize discussion about how to structure mandates.

E-REGS is inherently limited in that it does not, on its own, motivate action; its use presumes that federal actions take place. Contrast I/M with climate change, for example. During the 1990s, EPA moved aggressively to implement an I/M policy that improved on existing state and local efforts. In contrast, during the 2000s, under the George W. Bush administration, the agency took few steps to implement a control-oriented federal climate change policy, while state and local agencies were more proactive in their efforts to initiate controls (Lutsey and Sperling 2008). In cases where agencies choose not to act, E-REGS offers little

The Environmental Regulatory Spectrum (E-REGS)

| Performance standards | Command-and-control |
|---|---|
| **Politics** | **Politics** |
| • Weak economy<br>• Transition period between political administrations<br>• Politically strong stakeholders<br>• Preferred policy lacks political support | • Strong economy<br>• No political administration transition period<br>• Politically strong promulgating agency<br>• Preferred policy is a politically acceptable solution |
| **Problems** | **Problems** |
| • Evolving science for problem and solutions (lack of technical consensus)<br>• Variability in problems among jurisdictions | • Established understanding of problem and solutions (technical consensus)<br>• Uniform problems across multiple states or jurisdictions |
| **Policies** | **Policies** |
| • Public role necessary to implement solutions (behavior-oriented controls)<br>• Local or state implementation<br>• Implementing agencies have demonstrated record of successful past implementation experiences (e.g., states with strong environmental track records)<br>• Implementation outcomes less assured (pilot studies or full-scale implementation experience lacking) | • Limited public role necessary to implement solutions (technology-oriented controls)<br>• Federal implementation<br>• Implementing agencies have limited or nonexistent record of successful past implementation experiences (e.g., states with weak environmental track records)<br>• Implementation outcomes more assured (pilot studies or full-scale implementation results available) |

Performance standards ←→ Command-and-control

**Figure 10-4. Conditions that Favor Different Regulatory Management Approaches**

assistance. Although E-REGS can help identify conditions that merit action, it is primarily focused on motivating how, not whether, to pursue a policy response.

## Using E-REGS

With any guide such as the tool offered here, it is fair to ask how it might be used. After all, it is somewhat implausible to assume that a copy of E-REGS would hang in the EPA administrator's office, visible to all during discussions of pending policy proposals. The most straightforward response is to state that E-REGS is meant to stand as a thought-provoking set of considerations for professionals in the environmental management arena. It is my hope that those who read and digest the elements of E-REGS will retain an understanding of its essential message: that regulatory structure involves multidimensional considerations that can fundamentally affect the choice of a regulation's form. More than anything, the tool is meant to stimulate discussion among students and practitioners about how to design particular policies and ensure that an appropriate level of consideration has been given to the key factors that influence whether to lean toward prescriptive or more flexible mandates.

As a hypothetical illustration, consider how EPA participants involved in setting national I/M policy might have benefited if their policy-setting discussions included debate over the topics contained in E-REGS. Suppose that before the Smog Check debate period, an internal EPA peer-review discussion took place about the content and structure of the I/M policy being drafted by staff in Ann Arbor, Michigan (OMS). Assume the discussion addressed criteria in E-REGS (Figure 10-4). Also assume that discussion participants included scientists from several EPA offices, such Michigan, Nevada, and North Carolina (ORD). Consideration of the policy-setting factors included in E-REGS could have raised awareness about the pros and cons of RSD and IM240. As noted earlier, EPA scientists in Nevada and North Carolina held differing views from OMS scientists regarding RSD and IM240. ORD might have argued successfully to allow states flexibility to experiment with RSD, for example, noting the tremendous interest in the new technology while acknowledging uncertainties about its effectiveness (considerations identified by E-REGS). Thus an E-REGS-influenced discussion might have revealed the divisions within EPA on the RSD issue and raised warning flags for the political appointees to examine the issue further before establishing command-and-control regulations that effectively eliminated implementation options sought by states.

Many peer-review mechanisms have long existed to assist EPA managers in establishing policy. However, peer-review processes are often geared toward assessing a policy's technical merits. As the I/M case study illustrates and the E-REGS framework explicitly includes, the environmental manager also needs to assess other factors, such as a policy's political merits. Returning to our hypothetical peer-review discussion at EPA, suppose that after internal EPA debate, an I/M policy began to emerge. Consideration of the factors included in E-REGS could have prompted discussion about the policy's political merits. Appointees could have sought some evidence of the likely political reaction to the preferred option under consideration.

Another way E-REGS might have helped avoid the Smog Check debate is by enabling peer reviewers to explore expected state program adoption and implementation issues. For example, the discussion might have focused on whether policy implementation required approval by state legislatures. That may have helped appointed officials anticipate the extent to which political bargaining was going to take place. Political appointees might then have encouraged staff to obtain further external feedback on the political acceptability of the candidate strategy.

The hypothetical peer-review example merely serves to illustrate how the ideas embodied in E-REGS could promote useful policymaking discussions. It is important to emphasize that the example is not meant to overrepresent the usefulness of internal peer review as opposed to the involvement of external peer reviewers. There are limits to the effectiveness of peer review conducted solely within an organization. For example, peer pressure to be a team player could discourage peer reviewers from disagreeing too emphatically with their organizational counterparts, especially given the likelihood that at some point in the future, the roles for those peers would be reversed when their own prospective policies were subject to review. Therefore, although this example focused on an internal peer review, external peer review is also essential to the process of reviewing policy proposals. As the U.S. National Research Council articulated, broad-based deliberation involving important actors from outside the agency needs to take place prior to a major policy decision (NRC Committee on Risk Characterization 1996).

## CLOSING THOUGHTS

A school of thought in the policy analysis community holds that perhaps at least 10 years must pass before a policy can be fully evaluated. Various stakeholders learn over time which elements of a policy work best to resolve problems; this learning leads to continuous policy evolution. Policy critiques that occur too early may prematurely judge a policy's effectiveness (e.g., Lindblom 1959; Sabatier 1988; Sabatier et al. 1993). The U.S. I/M policy story, having evolved for many years following EPA's 1992 enhanced I/M regulations, was ripe for analysis.

As the U.S. political pendulum has swung back and forth over the years, a tendency among national political leaders has been to publicly favor deregulation or regulation. As a general characterization, these preferences have followed partisan affiliation, with more conservative political leaders leaning toward deregulation and more liberal representatives favoring regulation. The Smog Check story illustrates that those debating policy should not necessarily focus on deregulation versus regulation, but on designing and implementing the right types of regulatory approaches. Complicated environmental problems require government attention; the form that attention should take varies depending on the problems, the politics, and the policy options available.

With Smog Check and other U.S. I/M programs, time allowed states (if they chose) to correct program deficiencies. Stakeholders had the ability to learn and adapt, and policies evolved accordingly—although results varied by state. For a wide range of reversible environmental problems, the positive implication of the learning that takes place is that if mistakes are made, they will eventually be found out, and opportunities will exist to address them. Substantial time may pass during the learning process. Indeed, researcher Paul Sabatier, for example, described the policy evaluation time frame to be 10 to 20 years (1986, *39*). However, as stakeholders gather data and understanding grows about problems and policy options, a consensus eventually emerges about the effectiveness of various solutions.

An important caution needs to be raised here: not all environmental problems are reversible. Wilderness and habitat loss, endangered species extinction, and global climate change are examples of irreversible environmental problems. Very few windows of opportunity are likely to be available in which to address such problems. Thus, for irreversible cases, Smog Check's lessons—the importance of designing appropriate regulations—become even more important, as policymakers will have little time to correct mistakes. Smog Check ultimately serves as a reminder to create and implement scientifically sound, and politically acceptable, environmental management solutions.

## NOTES

[1] For a discussion of the most-cited Supreme Court decisions, see Santa Clara University 2008.

[2] As of 2009, EPA had established a deputy assistant administrator for science, based at the agency's Office of Research and Development.

[3] The concept that competition among states encourages reduced environmental protection is sometimes referred to as the "race-to-the-bottom hypothesis" (see, for example, Adelman and Engel 2008, *1804–1805*).

[4] Even within a given state, wide variability can exist regarding enforcement of environmental regulations. A 1990s study, for example, examined enforcement efforts against California oil refineries violating 1990 CAAA requirements. The study found a "vast discrepancy in both the rates of violations and the fines imposed on refineries in northern and southern California"; San Francisco Bay Area refineries received an average $700 fine per violation, whereas Los Angeles area refineries received an average $19,000 fine per violation (Hileman 1999).

CHAPTER 11

# EPILOGUE: VIEWPOINTS FROM OTHERS AT EPA

This chapter shares observations from David Howekamp, Felicia Marcus, and Dick Wilson, each of whom played an important role in resolving the Smog Check debate presented here. This discussion, along with the book's foreword by Mary Nichols, offers insights from former senior EPA air quality officials. These insights represent EPA views from the region and headquarters, as well as from career and appointed officials who were at the agency during the Smog Check conflict, and highlight the varied challenges faced by those serving in different roles and settings. Their contributions give readers an opportunity to hear alternative views and compare them with those expressed in earlier chapters.

Preparation of this epilogue occurred during 2007 and 2008, when I shared draft book materials with my former colleagues, and they in turn prepared written comments with their own Smog Check insights. I then worked with each person to consolidate his or her text, and finally, I integrated material from all the contributors to group common themes together. Each had final say over his or her text before its inclusion in this book.

The Smog Check conflict involved many stakeholders, each of whom may have a unique perspective. State officials in particular likely have important insights distinct from those presented here. I also shared the manuscript with state colleagues and sought their feedback, but as I believe the conflict ultimately originated from actions taken primarily by EPA actors, this chapter gives other EPA perspectives. Future work could more fully investigate viewpoints and insights from others.

David Howekamp served for more than 30 years in the federal government, leading the U.S. EPA air quality program in the southwestern United States (Region 9) from 1982 to 2000. While he was Region 9's lead air quality official, he oversaw Clean Air Act implementation in the worst-polluted U.S. region, managed a $40 million budget, and led 120 scientists, planners, and engineers. Following his EPA service, he provided air pollution consulting assistance to regional agencies, electric utilities, and private industry. His consulting work included service to environmental organizations as an expert witness in lawsuits, service on a Port of Los Angeles air quality advisory committee, and work to prepare a clean air plan for several Northern California cities, among other projects.

Before joining EPA, Felicia Marcus led the Los Angeles Department of Public Works after having been a private and public interest lawyer and community organizer in the Los Angeles environmental movement. She also served on the board of the Coalition for Clean Air, which, during her tenure with them, sued EPA to implement controls to meet the ozone air quality standards in Los Angeles, Sacramento, and Ventura. Additionally, she represented nonprofit environmental organizations before the South Coast Air Quality Management District Hearing Board. From 1993 to 2001, she served as President Clinton's appointed EPA Region 9 administrator. Following her service at EPA, she went on to become the executive vice president and chief operating officer of the Trust for Public Land, a nonprofit organization that conserves land across the United States, and later served as the western director for the Natural Resources Defense Council, an environmental organization.

Richard (Dick) Wilson served with the EPA for 31 years. During his tenure, he headed the Office of Mobile Sources (later the Office of Transportation and Air Quality) from 1982 to 1994, where he was responsible for overseeing all of EPA's transportation-related air pollution control programs. He became the deputy to Assistant Administrator Mary Nichols and eventually served as EPA's acting assistant administrator for air and radiation from 1997 to 1998. Following his retirement from government service, Dick worked as a senior vice president for the lobbying firm National Environmental Strategies. His work involved assistance to a wide array of clients, including Michigan State University, Sunoco Inc., and Toyota Motor Corporation.

## BACKGROUND TO THE DISPUTE

**Dave**: "I had a bittersweet experience in reading over the story once again. Smog Check was, in many ways, unique. When I compare it to the numerous issues I worked on at EPA, it didn't follow the typical pattern of how we worked with the states. My experience was that to get a new program adopted, we would usually work closely with the state air quality agency to agree on the program characteristics and needed legislation, and then to testify to the legislature in support of the program. That was the classic process I participated in (with mostly positive results) countless times. In this case, as I remember, the California Air Resources Board (CARB) was heavily involved up to the point of introducing Presley's original bill, the one with the Gold Shield stations. However, once EPA, meaning the Office of Mobile Sources (OMS), opposed the state's position, CARB basically cut their losses, and we were on our own."

**Dick**: "To understand how EPA established the enhanced I/M requirements, remember that in the early 1990s, car emissions were a major contributor to urban air pollution. Also, at the time Congress passed the 1990 Clean Air Act Amendments (CAAA), lawmakers were unhappy that EPA and the states had not made more progress toward clean air. Therefore, Congress set tough new requirements with short deadlines. One of the act's many new mandates was for stringent new-vehicle emissions standards. However, standards would take years to show benefits because of the need to wait for fleet turnover. That left better I/M as one of the few programs states could adopt to meet the act's near-term deadlines.

"Remember too that enhanced I/M was one of several controversial programs Congress included in the 1990 CAAA; it was by no means the only one that faced political pressure. As a result, EPA was implementing many difficult programs in the early 1990s. For example, in addition to I/M, the Ann Arbor–based team was also busy implementing new cleaner gasoline programs. EPA set these new clean-fuel requirements via an innovative negotiated rulemaking involving industry, states, environmental groups, and others—a first in EPA's history. Another innovative program involved acid rain. Acid rain controls used a creative emissions trading scheme—the first of its kind for a national program—that dramatically lowered control costs while accelerating emission reductions. The trading program became a world model and a template for early greenhouse gas reduction efforts. These programs illustrate the innovation EPA career staff employed when implementing the new act."

## SEEDS OF THE CONFLICT: FACTORS CONTRIBUTING TO THE DISPUTE

**Felicia**: "The two biggest mistakes the agency made were pushing test-only when not strictly required by the Clean Air Act (and not being clear that test-only was simply an EPA preference) and relying on the 50 percent discount without better supporting data.

"I think the timing issue was really important. Had the appointees been in place earlier, particularly Mary, with her depth of experience both in regulation and implementation from her years at the California Air Resources Board, we would have been more likely to avert the train wreck not just in California, but nationally. By the time Mary and I came in, however, there were so many regulations already on the conveyor belt that we didn't have time to properly evaluate and reconsider this one. Being good supportive soldiers, we initially carried on with the test-only mandate, trying to support our staff. That turned out to be perilous and ill advised.

"Another important contributor to the conflict was the way in which EPA (national) communicated with California. There were times when we acted like pointy-headed bureaucrats, versus people trying to solve a vexing environmental problem. Sometimes I felt like the points on our heads went through the ceiling tiles! California viewed our headquarters staff who had come before us as arrogant, three-piece suited, and nonresponsive to the state's needs. Overcoming this impression was initially the biggest part of my job [see below]."

**Dave**: "A contributing factor to the Smog Check dispute was the lack of personal accountability in a bureaucracy like EPA—no one person was accountable for the 50 percent discount. If a single person had to defend the agency's position, maybe we would have seen a different result. In a large bureaucracy, a 'group think' takes place as many individuals contribute to the shaping of a policy or decision. While that builds consensus and brings different ideas together, it waters down personal accountability. But it reflects the diffuse nature of programmatic responsibility across a big agency.

"For example, when I signed official EPA comment letters, such as some of those that went out during the Smog Check episode, the letters were shaped by numerous players. It was not unusual for official correspondence to be reviewed across all EPA regions, our headquarters office, and even our legal staff. By the time my signature went on a letter, I was acting more as a designated official signing for the agency as a whole rather than as an individual with personal accountability for the contents. If the process enforced more personal accountability, during Smog Check I probably would have asked tougher questions or dug more into the supporting material myself. But in a bureaucracy, an administrator relies on many people. My EPA experiences are in strong contrast to my life now as a consultant. Now, when I sign an expert report, I am held personally accountable for the findings—and that means I have to ensure the technical underpinnings are defensible. That's quite different from EPA's large-agency culture of diffused responsibility."

## INTERAGENCY DYNAMICS

**Dave**: "Smog Check happened right when some institutional friction surfaced between the career staff at CARB and EPA's Ann Arbor team (OMS). Usually the two organizations worked well together. During the early 1990s, however, there was an atmosphere of heightened tension because CARB had to get EPA approval on numerous efforts mandated by the Clean Air Act. One of the sore spots was estimating pollution from cars. OMS and CARB each used different computer models to estimate automotive emissions. As California assembled its air quality management plans, EPA and CARB had some pretty heated technical debates over the modeling assumptions being used. Those debates became more intense as studies emerged that showed the models underpredicted emissions. Smog Check came along in the middle of all this; it didn't help that EPA's I/M emission credits were based on an EPA computer model that CARB didn't use in California.

"In California, the Bureau of Automotive Repair (BAR) ran Smog Check, rather than the state air agency, and that contributed to some of the problems we experienced. EPA's institutional ties were strongest to our state air agency counterpart—CARB. BAR, by design, worked more closely with the repair industry and needed to remain sensitive to that constituency's needs. During the Smog Check episode, it felt a bit like BAR was always one step removed from the air quality improvement mission that drove CARB and EPA. Although we collaborated with BAR, we didn't achieve the same partnership with them that we had with CARB."

## THE NEED FOR POLITICAL AND PRACTICAL INPUT

**Felicia**: "EPA should have worked much earlier *and collaboratively* with California to improve I/M, especially given the state's long history with air quality control and its advanced regulatory framework. We should have done early and more effective vetting of our policy for customer/client friendliness. The whole controversy would not have happened if EPA had more respect for what it took to implement a program on the ground. EPA's *theory* made sense *theoretically*, but folks lacked the implementation experience to recognize how hard it would be to put our plan into practice—this was a key part of the problem, and it speaks to the need for getting appointees in place early. Political appointees often come in with more real-world practical experience, particularly in implementation.

"This is a great case study for how not to impose regulations! In the end, Ann Arbor's nightmares were of their own making. They should have anticipated California's reaction. OMS was naïve to the extreme about implementation and callous as to the costs. Despite my critiques of OMS, I do not question that they were well intentioned and trying to do the right thing. I just think a more pragmatic view of the world gets you further faster and less painfully."

## EPA'S I/M REGULATIONS

**Dick**: "While deciding how to define enhanced I/M, EPA was under pressure from the Clean Air Act, from states, from industry and others to achieve maximum emission reductions in the shortest time possible. EPA had to reach those goals with the information at hand, given the act's deadlines. We did not have time for more research or pilot studies. Did EPA have all the answers? Of course not. But our technical folks knew we needed a more sophisticated test. That is why Ann Arbor developed a short dynamometer test (IM240) based on EPA new-vehicle certification protocols. Also important was the experience of EPA and various states that test-only programs had worked much better than test-and-repair.

"No one at EPA thought the IM240 test was the only one that would work, or that test-and-repair programs couldn't be improved. States, however, were pressuring EPA to be tough—basically saying the only way to put good I/M programs in place was for EPA to set tough standards forcing dynamometer tests in test-only regimes. That view was supported by environmental groups, industry, and others. As a result, in 1992, near the end of the first Bush administration, the political leadership at EPA, including Administrator Bill Reilly and Assistant Administrator Bill Rosenberg, decided to take a tough position on the enhanced I/M definition. Another factor influencing the decision was that the Bush administration was committed to vigorously implementing the new act, since they had negotiated its passage with Congress.

"During the process to define enhanced I/M, career staff gave EPA's political leaders various options. Frankly, we were skeptical about EPA's ability to force test-only dynamometer programs. We knew that the vehicle service industry—test-and-repair program operators—would fight any major change and would have significant political power at the local and state levels. Nevertheless, once EPA's political leadership adopted a tough enhanced I/M definition, it became the career staff's job to implement and defend that position despite our reservations."

## SANCTIONS AND HISTORY

**Dave**: "The sanctions threat was, of course, highly contentious and contributed to the difficult tension that existed between the state and federal governments. There's a very important historical aspect to sanctions that sheds light on why the agency threatened their use during Smog Check. Simply put, sanctions had worked before. EPA imposed highway sanctions on California from 1980 through 1982 when the state failed to adopt its first I/M program. They were lifted once state lawmakers adopted authorizing legislation. So, interestingly, the Smog Check episode depicted here started out as a repeat of the same situation that had occurred more than a decade earlier. A key difference was that the first time, sanctions worked! EPA imposed sanctions in December 1980—the waning days of the Carter administration—just about a month prior to Reagan's inauguration as president. The careerists in that case were able to keep the sanctions in place even during the controversial Reagan–Gorsuch era in EPA's history. Once the sanctions were imposed, the pain didn't really hit for about two years; so many highway projects were already in the pipeline that it took a while for the funding lapse to have an impact. It was during that sanctions window that EPA helped coax the California legislature to adopt the first iteration of its Smog Check program."

## CALIFORNIA'S LEADERSHIP IN THE AIR QUALITY ARENA

**Felicia**: "A real frustration of mine was the tension that existed between granting California flexibility and the precedent it would set for other states. Historically, there have been plenty of periods where California has led the nation and the feds have followed, sometimes reluctantly. Given the state's preeminence in the air quality field, I would have preferred to have walled off California as a special case. There was certainly precedent for that in the Clean Air Act. Instead, we lumped California in with everyone else, and in the process set ourselves up for the 'domino' effect that occurred across the country after we gave California flexibility. The result: a perfect storm for us, to be sure."

## IMPLEMENTATION

**Dick**: "EPA launched the enhanced I/M requirements and, as expected, met a lot of pressure. Many states wanted the emission reductions such programs would achieve and moved ahead in combination with EPA to get necessary state legislation. Every state had its own set of problems. Once states moved to a test-only type program, they put pressure on EPA to not back down in any other state—since they knew one crack in the dike would end the whole effort. Unfortunately, all those folks who pushed EPA to set a tough enhanced I/M definition and who insisted EPA stick to that definition were largely absent in the public arena when the problems arose."

## SETTING PRECEDENT

**Dave**: "We can't forget a huge reason why EPA was so reluctant to cut California some slack: it was the fear of what would happen in other states. Most other states lacked California's air quality expertise and political will (most of the time). Yet if EPA granted California flexibility, we knew other states would want the same flexibility, regardless of their track record advancing clean air. We were very concerned that some states might abuse that flexibility and implement marginally effective programs. That was certainly an underlying concern of mine—I didn't want our actions in California to undercut the hard-won progress my counterparts had already achieved with their states. Remember that by the time the Smog Check dispute took place, several states had already committed to the EPA model program. We didn't want to lose those commitments. Unfortunately, reality bore those fears out: EPA's actions in California triggered a domino effect on I/M program commitments across the country."

## THE NEED FOR SOUND SCIENCE

**Felicia**: "Looking back, it seems outrageous to me that EPA based a multimillion if not billion dollar imposition on states without better data. That California called us on that is not surprising; that other states did not is. Requiring California to retool their entire Smog Check industry, with so little data to back up that mandate, was the lion's share of the problem.

"It helps to contrast what happened during Smog Check to what happened on two other California issues that took place simultaneously: the Bay-Delta water allocation deal and the court-ordered federal implementation plans, or FIPs. The Bay-Delta issue involved untangling decades of water-use wars that had pitted environmentalists against residential and agricultural water consumers. Yet we wrangled Governor Pete Wilson's administration into a great deal on Bay-Delta because we made the case to key stakeholders and the business community that a deal was in everyone's interest—and, this is key, we had virtually *no* scientific vulnerabilities in our argument.

"The FIPs were an outgrowth of the Clean Air Act's structure. California had failed to adequately achieve clean air in Los Angeles, Sacramento, and Ventura. Environmentalists, myself included, sued EPA to enforce the preparation and implementation of each region's air quality management plan (SIP). As a result, the courts ordered EPA to prepare and implement its own air quality plans (the FIPs) for those areas. We did pretty well with the FIPs, given the enormity of the task. Imagine, turning over the worst air pollution problems in the country to just a handful of EPA staff. With the FIPs, we did our scientific homework, plus we could make the case that we were stepping in where the state had failed, and we were at least trying to give folks the public health protection they deserved. Where we went overboard on the draft, which happened, we changed the final to reflect what we heard. Ultimately, we didn't have to impose the FIPs, but we did leverage an approvable SIP, which is the ultimate intent of the Clean Air Act.

"Each issue—Smog Check, Bay-Delta, FIPs—was hugely complicated. Overall, we fared much better when we had a stronger legal or regulatory foundation to support our actions."

**Dave**: "Smog Check is a good example of the need to improve scientific understanding when setting policy; the lack of adequate science was a huge flaw on EPA's part. The lack of adequate science in EPA decisionmaking is still a major challenge that remains difficult to address even today, especially given the mismatch between EPA's numerous statutory obligations and the financial and staff resources available.

"As we think about the broader lessons from this experience, those that extend beyond Smog Check, we must be cautious, though, of expecting good science to be a panacea. There are too many instances where the policy-setting process has ignored science or, even worse, when the science was manipulated or watered down. For example, in the early 2000s, we saw political manipulation to downplay scientific warnings about global warming. Another egregious example, from the mid-2000s, was EPA's rejection of scientific advice on setting National Ambient Air Quality Standards. So while we need to pursue sound science to support decisionmaking, we need to be realistic about its application in the political arena."

## I/M AS A TOOL TO ACHIEVE CLEAN AIR

**Felicia**: "Despite all of the problems with EPA's position on test-only, the agency did get the big picture right about the need to improve I/M. Some form of effective Smog Check improvement was, quite simply, one of the most cost-effective approaches to achieve clean air. Cost per ton reduction for Smog Check is a fraction of stationary source reduction. Enhancing I/M was a no-brainer and the right thing to do as a policy call. How we sought to implement it was the problem."

## NEGOTIATIONS: REBUILDING RELATIONSHIPS

**Felicia**: "To get to a deal, I needed to know what our real 'interests' were, versus our 'positions,' and I needed to understand the same things for the state. That's how you get to real resolution. I realized early on that the best way I could help EPA, and achieve cleaner air, wasn't to blindly charge forward mouthing the agency's standard talking points about I/M. It was to add that little 'p,' political experience, meaning political savvy that is not partisan. That's the way the legislative world works. In fact, it is the only way anywhere to get things to work, truly. Onesize-fits-all should only be applied if it really does fit.

"It was so important to get past the posturing and to have conversations where each side heard each other. In my case, I knew many of the key players, and we genuinely liked each other, and I knew they trusted *me* as an individual. But it wasn't enough to just talk to them; I had to acknowledge their point of view, where we had been in the dispute, and where we were now. I had to get them to a point of feeling heard, respected, and in a position to get somewhere in the future. In the closing months of the Smog Check debate, which is where I came in, I knew, from meeting individually with the legislative and state people, how they viewed EPA. We certainly had lost all leverage as a player in their world.

"One of the biggest contributions I was able to make was to help the state participants see what was in it for them to reach a deal and get past the endless negotiating. I spent hours in private one-on-one discussions with senators, assemblymen, and their staff, rebuilding a rapport just to get them willing to talk to us again. I can now share a story from that time that brings a smile to hear it but, when it occurred, was very telling about how deep a hole EPA appeared to be in when I arrived. During a lot of my one-on-one conversations, I focused on explaining, convincing really, that our I/M regulations did *not* actually require 100 percent separation of test and repair—rather, the regulations required states to come up with an *equivalent* program. At one point, I had explained this to Senator Newton Russell twice, to which he had disagreed with me both times, saying 'No. EPA requires separation.' I finally pulled out my EPA ID badge, saying, 'No, really, Senator. I swear to God. I now work at EPA, and really, we *prefer* separation of test from repair for a host of reasons, but it is not absolutely required.' Finally, he sat back, looked at me and my ID card, and said: 'Well, then . . . ' And that was what gave us the opening to actually negotiate. He could believe that Felicia wouldn't require it, but not that EPA wouldn't, because of what he had been told not just by EPA, but by the Wilson administration in their attempt to discredit the agency.

"A key lesson from that experience, and I know Senator Russell has shared this story as well, is that *people hear what they expect to hear rather than what you actually say.* As I look back upon my private conversations with Russell and the others, I think those hours were my real contribution to making it even possible for us to discuss a deal.

"Similarly, as Doug notes in a passage that made me laugh so hard I was crying when I read it, I knew that I had to talk long enough during one negotiation that they would eventually really hear *me*. I started with the history of us not listening to them, acknowledged every one of their arguments, and then made the case for why it was worth listening to us and coming back to the table now. I knew that as politicians, and as aggrieved people who felt disrespected, their instinct would be that they could 'win' the battle with us. I had to bring them back to thinking about air quality, their legacy, and the opportunity to 'succeed' at something. And I had to keep trying until

I could see that they were seeing 'Felicia' and hearing me versus being in a fugue about 'EPA.'"

## ROLES OF THE PRESS AND THE PUBLIC

**Felicia**: "I thought we did better with the press than most agencies. I personally felt I had the chance to accomplish quite a bit with my press contacts; I knew from experience that if the press trusted you, you would get a better story. Spending time on background is important, as is thinking about language. We couldn't stop anyone from using a great biting quote against us, but we could mute the damage. And we did. As well as get a little credit for trying to do the right thing. And, most important, getting the facts out in a sea of political rhetoric is challenging but doable. Our biggest problem was the facts over time: EPA's assertions lacked the scientific data needed to support them, and eventually the press understood this—as did we!

"Earlier chapters in this book suggest that some of the problems experienced might have been averted through more active public participation, especially during implementation. At least during the negotiations, I'm not sure how public engagement would have helped. State politics in California is intensely personal, and the key negotiators were the ones who had the power to craft a Smog Check agreement. Some early hearings might have presaged challenges to come from key stakeholders, like the service station owners, however, and that would have been extremely helpful."

## SOLUTIONS

**Felicia**: "We would have gotten further faster by agreeing to an 'amped up' Gold Shield program, especially since we knew cleaner, OBD-equipped cars were coming. Instead of requiring 'our' program, we should have accepted 'their' Gold Shield program, but with added bells and whistles to improve its effectiveness. In other words, we would have gotten further faster by saying yes to something, asking for a little more, and then getting to 'trust but verify' or 'show me the emission reductions.' Ironically, after all the pain, that is where we ended up anyway."

**Dave**: "Upon reflection, I wish I would have pushed back against OMS positions early on, rather than accepting their viewpoints. It should have been more of a red flag for us in the regional office when others in California, who were knowledgeable about I/M, were skeptical of the technical basis for OMS's position. Those of us in the field assumed too readily that the experts at OMS had a strong scientific case for their position. Perhaps we in the regional office were overly skeptical of viewpoints opposed to EPA because it seemed that so much of the pressure against us came from those who had a vested interest in retaining the existing Smog Check program."

## POLITICS: PRESSURE FROM CONGRESS

**Dick**: "Due to the 1990 CAAA's aggressive implementation deadlines, political pressure against the act began building on many fronts soon after President Clinton's 1992 election and peaked with the Republican takeover of Congress two years later. EPA administrator Carol Browner and assistant administrator Mary Nichols, President Clinton's replacements for Bill Reilly and Bill Rosenberg, found themselves in difficult times; the California I/M episode was one example of many challenging issues. Around the time of the Republican takeover in Congress, I shifted up a level to be Mary's deputy (the top career job in the air office) and worked closely with Mary and Carol and the EPA staff to develop ways to ease up a bit on almost every one of the new Clean Air Act programs. Letting some steam out of the system was a political necessity, and we found we were mostly able to keep the basic benefits while giving some on time and flexibility.

"My own sense is that if Carol and Mary switched places with Bill and Bill, the same decisions would likely have been made in the same time frames. All four appointees were dedicated public servants who cared about EPA and the environment and who operated well in the political environment they were given."

## PROBLEM-SOLVING OWNERSHIP

**Felicia**: "Something very important is the need for people to take 'ownership' for their actions. Once EPA gave California the flexibility to implement the program they wanted to, the state took ownership over that effort. When California experienced implementation problems in those initial months, recall that they did not blame EPA for the foul-ups: they had given us 'their' plan, and the implementation problems were theirs as well. My view was that it was likely that implementation would be hard to do, whatever plan we agreed to. However, had they grudgingly (or under coercion) accepted our full test-only mandate, they would have then blamed us for the implementation problems that would have inevitably resulted. Had test-only been absolutely the right answer, I would have pushed to go with it and counseled that we stay engaged on the implementation. By giving California a deal they owned, however, they had a far greater incentive to make it work, and they owned the problems if it didn't. That's an important lesson that cuts across a lot of environmental (and other policy) issues."

## A COMMENT ABOUT CAROL BROWNER

**Felicia**: "Although some of the early press reports characterized Carol Browner as a relative unknown in the environmental arena, I think that was unfair. She might have been less well known to some in the western United States, but she was not an unknown to the environmental world. When Al Gore was a U.S. senator, Browner was his legislative aide, and she also served as secretary of Florida's Department of Environmental Regulation, one of the nation's largest state environmental agencies. Time showed that she was brilliant, as green as they come, and she got a lot done. I feel she virtually single-handedly faced down the later hostile Congress and helped turn the president into a green advocate."

## PARTISAN POLITICS

**Felicia**: "Much of the debate was a state versus federal conflict that crossed party lines. Still, there were underlying partisan motivations that helped fuel the controversy, and they need to be acknowledged. During the conflict, California's Republican governor, Pete Wilson, was preparing to run for reelection, and at that time the election outcome was very uncertain. His administration took the opportunity to use Smog Check—and every other environmental regulatory action on air, water, and waste—as a political hammer to bash the Clinton Democratic EPA. He ran against the federal government, regulation, and immigrants. This was an undercurrent throughout the debate. For example, while the Wilson administration's fortitude in critiquing the 50 percent discount was accurate and justifiable, they prolonged debate over that topic to lengthen the dispute and, in my mind, score political points against the Democrats. It's one of the reasons Katz, who was a Democrat, grew so frustrated toward the end of the negotiations. He was trying to strike a deal, while the Wilson administration was posturing and setting up roadblocks. Frankly, we would have made a deal earlier had it been only us and the legislature."

## GETTING TO "GREEN"

**Felicia**: "Some who read this story may think that EPA career staff were more 'green' than the EPA appointees who eventually compromised and struck a deal. Nothing could be further from the truth. That EPA's I/M position was wrong and couldn't hold up over time didn't mean the career staff were greener. Remember that many of us, meaning the appointees, had to file citizen suits against EPA earlier in our careers because the agency hadn't moved fast enough to protect the environment.

"*Each issue has to stand on its own merit.* Being tougher isn't necessarily being smarter, especially if one takes a broader view. The house of expensive regulations built by EPA over the years is what led to a lot of the revolt from the field. Figuring out how to get to the best result *at the least cost* is absolutely more 'green' in the long haul. You also have to have compassion and intelligence about the way these things play out on the ground. When I was in local government, I had to implement expensive regulations that didn't get the best environmental results, and I had to pass those costs on to small businesses and residents. So I admit that when I joined EPA, I shared California's skepticism about the wisdom of regulations crafted on high. Yet I'd challenge anyone to be greener than me, or to stack up actual environmental results on the ground. I don't say that to sound defensive, but to clarify: often the best results are achieved through less prescriptive, more creative, more collaborative approaches. If one is interested in long-term environmental progress, one has to think about cost and implementation and take responsibility for it."

## LESSONS

**Dave**: "The lack of any substantial support from the environmental community, and in fact, in some cases their outright opposition to our policy, should have been a wake-up call that something was amiss. With perhaps only one exception, involving permits for stationary source operations, I can't remember another significant air quality fight where we didn't have the environmental community pushing hard on us to support our program or one even tougher. That should have been a clue that this was a fight headed for disaster."

## STRUCTURING CONTROL PROGRAMS

**Dick**: "An important lesson from the I/M story is the difficulty of implementing what I'll call *retail* versus *wholesale* programs. By wholesale, I mean programs adopted and implemented at the federal level. Examples are new-vehicle emission standards, national fuel standards, and the acid rain program. Wholesale programs delivered major clean air benefits with minimum government resources and political hassle. They have historically been our most productive clean air activities. I/M, on the other hand, provides a good example of the problems with retail programs. Retail programs require each state to pass legislation or regulations and to make implementation decisions under some EPA guidance and oversight. If the program is at all controversial, which I/M certainly was, then each jurisdiction has to deal with all of the different interest groups in their state. Complicating matters is that states generally do not publicly welcome federal standards, guidance, or oversight—even if required by Congress. Also, the states very much watch what each other is doing, and that makes it difficult for one state to be tougher than other states. You get into the impossible dilemma of states telling EPA privately that they want EPA to take a tough stand—tough standards, use of sanctions, and so on—and to stick to that stand with all states, while publicly complaining about overzealous federal bureaucrats.

"The contrast between private encouragement for EPA to be resolute and public lambasting at EPA being too rigid certainly happened with I/M; that is what made it so difficult for the political leadership at EPA, both national and regional, to deal with California. Anything done to respond to California tended to undercut the progress in other states. That is what Carol Browner faced in deciding how to handle the California sanctions issue. Here I disagree with Doug's interpretation: the sanctions dilemma was not the outcome of career staff giving the administrator poor choices. It resulted from the administrator and the career staff facing an impossible political dilemma. Having been there, I know the staff provided the administrator with a full range of options and a straight assessment of the pros and cons of each.

"Luckily, over time the importance of I/M programs has diminished as other *wholesale* programs have made cars less polluting. EPA subsequently required the auto industry to make emission control systems durable for the life of the vehicle and to install, after pioneering work in California, on-board diagnostic (OBD) systems that alert owners, inspectors, and service mechanics if there is a problem affecting emission control systems. As a result, cars now require less maintenance, taking the pressure off I/M."

## GOVERNMENT REINVENTION

**Felicia**: Issues like Smog Check reinforce the need for the reinvention efforts begun under President Clinton. The fact that reinvention is difficult and takes years to accomplish should come as no surprise. From the beginning of my tenure with EPA, Carol Browner told me that we were going to be about management and fundamental change, which would take a long time. I think it was heroic that she took it on, especially knowing that we needed two terms to even make a dent. All significant changes like this can take a decade or more. Plus, reinvention really encompassed two things: working internally to reinvent the culture within EPA to be more focused on ultimate effectiveness versus being 'right'; and working externally with our tribal, state, local, private, and community partners to reinvent our partnerships with them.

"Working to reinvent the internal EPA culture was hard, but working to reinvent our relationship with the states was even harder. An example of where we made some progress involves Indian tribes in the western United States. While at EPA, I worked to get more resources for tribes so they could build capacity to more effectively address their own problems. Part of why I felt comfortable giving tribes more resources and autonomy was that I felt we at EPA were too focused on micromanaging states, who had advanced regulatory structures, which is what we had always done, while the agency had ignored the tribes, who needed our help to build their programs and protect their people's public health and natural resources. By empowering tribes, EPA prioritized work that would yield great and overdue public health and environmental protection and helped them do a much better job of delivering on their own top-priority public health issues. Ditto with local governments. Engaging business and community partners was just good government, in my book, and helped counter the agency's tendency to focus inward rather than toward the public we served."

## COMMAND-AND-CONTROL VS. PERFORMANCE-BASED PROGRAMS

**Dave**: "I'm not a huge fan of the argument that performance-based regulatory approaches are superior to command-and-control. I realize that I may be biased by being an ex-member of the bureaucracy, but command-and-control has contributed enormously to the tremendous air pollution control progress we've made to date. Much of my post-EPA consulting work has involved the control of diesel emissions at ports, and it's my observation that, some 40-plus years into the air quality control process, CARB is still adopting command-and-control regulations at a rapid pace, as are the ports themselves. Flexibility is included in many of these rules, which you might call performance-based regulation on the micro level. However, the fundamental underlying principle remains command-and-control. To a certain extent, this debate always gets down to definitions. One could say that the requirement to have an I/M program, regardless of its design, is a command-and-control mandate, while flexibility to design the I/M program provides a performance-based approach that focuses on outcomes.

"I think we need to be very careful in our interpretation of the I/M experience. We shouldn't use I/M to advocate for performance-based programs across the board. I/M is a unique case. It is neither fish nor fowl—it has elements of both technology-based control and behavior modification. I can see an argument where the behavior change elements of environmental management are better suited to performance-based regulation, while the technology-oriented elements are better suited to command-and-control mandates.

"Ironically, one could argue that the Clean Air Act sets up one of the biggest performance-based environmental programs. The act requires that states submit to EPA their state implementation plans, or SIPs. The SIPs, the air quality management plans for each region, are performance based; they must demonstrate how a region will attain the National Ambient Air Quality Standards. However, states have the freedom to choose whatever strategy they need to achieve attainment. The verdict is mixed on whether that approach has worked. For the most part, it has worked in California in those areas that have either achieved or made huge progress toward meeting the standards. In the rest of the country, the evidence is less compelling—many states have relied too heavily on federal control programs, such as new-vehicle emission standards, and have made only slow progress toward attainment."

## CONGRESSIONAL GOALS VS. REALITY

**Dick**: "In my view, the overarching lesson from I/M is that our system of environmental lawmaking is not optimum. We amend our environmental laws infrequently, and then set tough new requirements and deadlines that are next to impossible to meet. Then EPA is expected to implement those laws aggressively and to fix the resulting programs when the political pushback gets too great to ignore. Of course, Congress tends to blame EPA and not themselves for these dilemmas. And it is all made more complex when administrations change and when Congress changes hands.

"Perhaps Congress would have been wiser to rely more on the 'wholesale' programs I mentioned, like new-vehicle standards and cleaner fuels, and to have given EPA and the states more time to meet air quality goals. But that is not our system. Congress tends to ignore problems until they get so bad they require action, and then they tend to overreact by expecting the problem to be solved overnight. When that cannot be done, it is the bureaucrats at all levels who have to fix the problem, while taking some political heat along the way.

"I think that we probably made a mistake to spend so much time and political energy—and chips—on enhanced I/M. I also think the path we ended up on was pretty inevitable, given the Clean Air Act's mandates and deadlines. The 1990 CAAA put in place forces that led to EPA taking an overly aggressive view of enhanced I/M—the act, even more than decisions made by EPA's political or career staff, was what drove us to the outcomes described here.

"I/M is an excellent example of what can go wrong in implementing new congressional mandates. If looked at alone from just a regional perspective, however, one can easily reach the wrong conclusions as to what led to those problems and how they might have been avoided. During the I/M conflict described here, EPA was blessed with some of the best, most innovative career staff at both headquarters and the regions. We were also blessed with excellent political leadership during both the first Bush and Clinton administrations. It would be wrong and too simple to blame the I/M outcome on bureaucrats or politicians without understanding the important context in which the controversy occurred."

## ASKING THE RIGHT QUESTIONS

**Felicia**: "I'll close by emphasizing how important it is for an appointee, or any manager, to ask the right questions. As relayed in the story, the first time I was briefed on Smog Check, shortly after my EPA appointment, my reaction was, 'Why can't EPA do what Katz is saying and modify the current program?' It proved to be the right question to ask. I needed to understand what EPA's rationale was to see if it held up, and I needed to understand why we were in dispute. If EPA had fully answered that question early in the policy-setting process, Smog Check could have served as a very different case study."

# REFERENCES


Adelman, D.E., and K.H. Engel. 2008. Adaptive Federalism: The Case Against Reallocating Environmental Regulatory Authority. *Minnesota Law Rev.* 92(June): 1796–850.

ADEQ (Arizona Department of Environmental Quality). 2000. Responsiveness Summary to: Testimony Taken at the Ozone Oral Proceeding and Written Comments Received on the Serious Area Ozone State Implementation Plan for Maricopa County, December 14.

Air Pollution Control Association. 1982. *JAPCA* 32(1).

Allison, G.T. 1971. *Essence of Decision: Explaining the Cuban Missile Crisis.* Boston, MA: Little, Brown and Company.

Amlin, D. 2002. California's OBDII Program Update. Presented at the Colorado State University 18th Annual Mobile Sources/Clean Air Conference, Breckenridge, CO, September 10–13.

Arizona State Legislature. 1999. *Vehicle Emissions Inspection Meeting.* Meeting minutes, Legislative Study Committee, Arizona State Legislature, September 21. http://www.azleg.state.az.us/iminute/house/0921veip.htm. Accessed April 26, 2008.

Aroesty, J., G. Farnsworth, L. Galway, M. Kamins, L. Parker, D. Rubenson, and P. Wicinas. 1993. Restructuring Smog Check. Testimony to the California Senate Transportation Committee, August.

Asian Development Bank. 2003. Policy Guidelines for Reducing Vehicle Emissions in Asia: Vehicle Emissions Standards and Inspection and Maintenance, Publication No. 110602. Manila, Philippines.

Auer, M.R. 2008. Presidential Environmental Appointees in Comparative Perspective. *Public Admin. Rev.* 68(1): 68–80.

Bachman, J. 2007. Will the Circle be Unbroken: A History of the U.S. National Ambient Air Quality Standards. *J. Air & Waste Manag. Assoc.* 57: 652–97.

Bardach, E. 1996. *The Eight-step Path of Policy Analysis (a Handbook for Practice).* Berkeley, CA: Berkeley Academic Press.

Barrett, R.A., R.A. Ragazzi, and J.A. Sidebottom. 2005. *Colorado OBDII Vehicle Evaluation Study.* Final report. Denver, CO: Colorado Department of Public Health and Environment, December 20.

Barton, J. 1995. Letter from Joseph Barton, U.S. House of Representatives, to Mary Nichols, U.S. Environmental Protection Agency, Washington, DC: March 20.

Baugh, X. 1997. California State Assembly Bill No. 1492, Chapter 803: Air Pollution: Motor Vehicle Inspection and Maintenance, October 9.

Bertelli, A.M., and L.E. Lynn. 2006. *Madison's Managers: Public Administration and the Constitution.* Baltimore, MD: Johns Hopkins University Press.

Beusse, R., J. Bishop, M. Brown, D. Cofer, D. Howard, T. Johnson-Davis, P. Milligan, and B. Nelson. 2006. *EPA's Oversight of the Vehicle Inspection and Maintenance Program Needs Improvement.* Evaluation report. U.S. Environmental Protection Agency, Office of Inspector General, Research Triangle Park, NC: 2007-P-00001, October 5.

Bishop, G.A., D.H. Stedman, B.R. Hutton, L. Bohren, and N. Lacey. 2000. Drive-by Motor Vehicle Emissions: Immediate Feedback in Reducing Air Pollution. *Env. Sci. Technol.* 34(6): 1110–6.

Bishop, G.A., D.A. Burgard, and D.H. Stedman. 2006. Emissions from Vehicles Never Subject to I/M Programs. Presented at the 16th Coordinating Research Council On-Road Vehicle Emissions Workshop, San Diego, CA, March 28–30.

BLS (Bureau of Labor Statistics). 2003. Household Data Annual Averages: Employment Status of the Civilian Noninstitutional Population, 1940 to date. U.S. Department of Labor, Bureau of Labor Statistics. ftp://ftp.bls.gov/pub/special.requests/lf/aat1.txt. Accessed April 26, 2008.

Bowler, L. 1996. Assembly Bill No. 2515, Chapter 1088: Vehicle Inspection and Maintenance, September 29.

Boyd, J.D. 1992. Letter from James D. Boyd, California Air Resources Board, to Daniel McGovern, U.S. Environmental Protection Agency, Washington, DC: November 13.

Brown Jr, W.L., R. Katz, N.R. Russell, D. Boatwright, and Q.L. Kopp. 1993. California Legislature. Sacramento, CA. Letter to Carol M. Browner. U.S. Environmental Protection Agency, Washington, DC: September 10.

Browner, C.M. 1993a. U.S. Environmental Protection Agency. Washington, DC. Support of Senate Bill 119 by the U.S. Environmental Protection Agency. Letter to Robert Presley, Sacramento, CA: California State Senate, August 26.

———. 1993b. U.S. Environmental Protection Agency. Washington, DC. Letter to David Roberti. Sacramento, CA: California State Senate, September 10.

———. 1994a. U.S. Environmental Protection Agency. Letter to California governor Pete Wilson, Washington, DC: January 24.

———. 1994b. Letter from Carol Browner, Administrator, U.S. Environmental Protection Agency, to Jim Folsom, Jr., Governor, state of Alabama, December 20.

Browner, C.M., and F. Pena. 1993. U.S. Environmental Protection Agency and the Department of Transportation, Washington, DC. The U.S. Environmental Protection Agency urges California Governor Pete Wilson to Adopt Enhanced Vehicle Emissions Inspection and Maintenance Programs to Avoid Sanctions. Letter to Pete Wilson, Governor of California, Sacramento, CA, April 13.

Bryner, G.C. 1995. *Blue Skies, Green Politics: The Clean Air Act of 1990 and its Implementation.* 2nd ed., Washington, DC: CQ Press.

BTS (Bureau of Transportation Statistics). 2001. *National Transportation Statistics 2000*, BTS01-01, April. http://www.bts.gov/publications/nts/2000/index.html. Accessed April 26, 2008.

Burnette, A.D., S. Kishan, and T.H. DeFries. 2008. Evaluation of Remote Sensing for Improving California's Smog Check Program. Final report, version 15, prepared for the California Air Resources Board and the California Bureau of Automotive Repair, by Eastern Research Group, Inc., Austin, TX. ERG No. 0023.01.001.710, March 3.

Burstein, P. 2006. Why Estimates of the Impact of Public Opinion on Public Policy are Too High: Empirical and Theoretical Implications. *Social Forces* 84(4): 2273–89.

Bush, G.W. 1995. Letter from George W. Bush, governor of Texas, to Carol Browner, administrator, U.S. Environmental Protection Agency, April 12.

Cadle, S.H., P. Mulawa, E.C. Hunsanger, K. Nelson, R.A. Ragazzi, R. Barrett, G.L. Gallagher, D.R. Lawson, K.T. Knapp, and R. Snow. 1998a. Light-duty Motor Vehicle Exhaust Particulate Matter Measurement in the Denver, Colorado, Area. In *Proceedings of Air & Waste Management Association, $PM_{2.5}$ A Fine Particle Standard Conference.* Vol. II. Edited by J. Chow, P. Koutrakis, Long Beach, CA, January 28–30, pp. 539–58.

———. 1998b. Measurement of Exhaust Particulate Emissions from in-use Light-duty Motor Vehicles in the Denver, Colorado, Area. CRC Project E-24-1. Atlanta, GA: Final report prepared for the Coordinating Research Council, March.

CA DOF (California Department of Finance). 2008. July 2008 California Employment Highlights: Current Job Growth Trends in Comparison to Past Recessions, Sacramento.

CA IMRC (California Inspection and Maintenance Review Committee). 1993. Evaluation of the California Smog Check Program and Recommendations for Program Improvements. Fourth Report to the Legislature, February 16.

———. 2000. Smog Check II Evaluation. Sacramento, CA, June 19.

———. 2006. Review of the Smog Check Program, November 28.

———. 2007. Review of the Smog Check Program, November 27.

CalEPA (California Environmental Protection Agency). 1995. Strock Praises Smog Check Plan Study. Press release. Sacramento, CA, April 3.

California State Assembly. 1951. Final Summary Report of the Assembly Interim Committee on Air and Water Pollution. Report prepared for the California State Assembly, Sacramento, CA, June.

Calvert, J.G., J.B. Heywood, R.F. Sawyer, and J.H. Seinfeld. 1993. Achieving Acceptable Air Quality: Some Reflections on Controlling Vehicle Emissions. *Science* 261: 37–45.

CARB (California Air Resources Board). 1999a. California Exhaust Emission Standards and Test Procedures for Model Year 2001 and Subsequent Model Passenger Cars, Light-duty Trucks, and Medium-duty Vehicles Adopted August 5, 1999. Sacramento, CA. http://www.arb.ca.gov/regact/levii/to_oal/ldvtp01.pdf. Accessed April 26, 2008.

———. 1999b. California Exhaust Emission Standards and Test Procedures for 1998–2000 Model Passenger Cars, Light-duty Trucks, and Medium-duty Vehicles as Amended August 5, 1999. Sacramento, CA. http://www.arb.ca.gov/msprog/levprog/cleandoc/ldvtp88.pdf. Accessed April 26, 2008.

———. 1999c. LEV II: Amendments to California's Low-emission Vehicle Regulations. Fact Sheet Prepared by the California Air Resources Board, Sacramento, CA, February.

———. 2000a. California's Air Quality History: Key Events, April 21. http://www.arb.ca.gov/html/brochure/history.htm. Accessed April 26, 2008.

———. 2000b. *Final Evaluation of California's Enhanced Vehicle Inspection and Maintenance Program (Smog Check II).* Appendix A. Sacramento, CA, July 12.

———. 2000c. *Public Meeting to Consider Approval of Revisions to the State's on-road Motor Vehicle Emissions Inventory.* Technical support document prepared by the California Air Resources Board, Sacramento, CA, May.

———. 2001. The California Low Emission Vehicle Regulations as of May 30, 2001. http://www.arb.ca.gov/msprog/levprog/cleandoc/LEVRegs053001.pdf. Accessed April 26, 2008.

———. 2007. Proposed State Strategy for California's State Implementation Plan (SIP) for the New Federal $PM_{2.5}$ and 8-hour Ozone Standards. *Draft Statewide Air Quality Plan*, April 26.

CARB and DCA (California Air Resources Board and Department of Consumer Affairs). 2004. Evaluation of the California Enhanced Vehicle Inspection and Maintenance (Smog Check) Program. Draft report prepared for the Inspection and Maintenance Review Committee, April.

Carlisle, R. 2004. Personal communication with Rocky Carlisle, Executive Officer. California I/M Review Committee, Sacramento, CA, April 19.

CEDD (California Employment Development Department). 2003. Labor Market Information: California Seasonally Adjusted, Industry Employment & Labor Force by Month, Sacramento.

Chrysler Corporation. 1998. Emission and Fuel Economy Regulations; CIMS 482-00-71; Environmental and Energy Planning. Auburn Hills, MI, April.

CiREM (Cambridge Institute for Research, Education and Management). 2004. Website prepared by Cambridge Institute for Research, Education and Management, Cambridge, England. http://www.cirem.co.uk/definitions.html#c. Accessed December 28, 2009.

Cohen, M.D., J.G. March, and J.P. Olsen. 1972. A Garbage can Model of Organizational Choice. *Admin. Sci. Quarterly* 17(1): 1–25.

Cohen, R.E. 1995. *Washington at Work: Back Rooms and Clean Air.* 2nd ed., Needham Heights, MA: Allyn and Bacon Press.

Commission of the European Communities. 2001. *European Governance: A White Paper.* Brussels, Belgium, July 25.

*Dallas Morning News.* 1995. Bush Signs Bill to Delay Emissions Tests, February 1.

Davis, D. 2002. *When Smoke Ran Like Water.* New York: Basic Books.

Degobert, P. 1995. *Automobiles and Pollution.* Society of Automotive Engineers, Inc., Editions Technip, Warrendale, PA, and Paris, France.

DeMaio, C.D. 2001. A Report to the 43rd President and 107th Congress. Managing for Results at the U.S. Environmental Protection Agency. Bipartisan Observations and Recommendations to the New Administration and Congress on Improving Management and Performance of the U.S. Environmental Protection Agency, February 1. http://reason.org/news/show/managing-for-results-at-the-us. Accessed December 28, 2009.

Downs, A. 1967. *Inside bureaucracy.* Boston, MA: Little, Brown and Company.

Eastern Research Group, Inc. 2002. Baseline Assessment of the State of the Science for Alternative Technology Options. Final report prepared for the Arizona Department of Environmental Quality, Phoenix, AZ: June 28.

———. 2006. Arizona Alternative Compliance and Testing Study (AZACTS): Summary of Activities and Recommendations to Date. Phoenix, AZ: Final report prepared for the Arizona Department of Environmental Quality. ERG No. 0147.00.008.002, January.

Edwards, G.C., and B.D. Wood. 1999. Who Influences Whom? The President, Congress, and the Media. *American Political Sci. Rev.* 93(2): 327–44.

Eisinger, D.S. 2005. Evaluating Inspection and Maintenance Programs: A Policy-making Framework. *J. Air & Waste Manag. Assoc.* 55: 147–62.

Eisinger, D.S., and P. Wathern. 2008. Policy Evolution and Clean Air: The case of U.S. Motor Vehicle Inspection and Maintenance. *Transportation Research Part D: Transport and Environment* 13: 359–68.

Elliott, E.D. 2003. Strengthening Science's Voice at EPA. *Law & Contemp. Problems* 66: 45–62.

Energy Foundation. 2001. Bellagio Memorandum on Motor Vehicle Policy Principles for Vehicles and Fuels in Response to Global Environmental and Health Imperatives. Consensus Document. Bellagio, Italy, June 19–21.

Faiz, A., and P.J. Sturm. 2000. New Directions: Air Pollution and Road Traffic in Developing Countries. *Atmos. Env.* 34: 4745–6.

Fifer, B. 2002. *Discovering Lewis and Clark, Coal Takes Over.* Washburn, ND: Lewis & Clark Fort Mandan Foundation, February.

Frederickson, H.G. 1994. Review Essay: Some Policy Implications of the Federal Reinventing Government Report. *Policy Currents* 4(1): 1–3.

Gardetto, E., T. Bagian, and J. Lindner. 2005. High-mileage Study of On-board Diagnostic Emissions. *J. Air & Waste Manag. Assoc.* 55: 1480–6.

Gauderman, W.J., E. Avol, F. Gilliland, H. Vora, D. Thomas, K. Berhane, R. McConnell, N. Kunzli, F. Lurmann, E. Rappaport, H. Margolis, D. Bates, and J. Peters. 2004. The Effect of Air Pollution on Lung Development from 10 to 18 Years of Age. *New Engl. J. Med.* 351(11): 1057–67.

Gauderman, W.J., H. Vora, R. McConnell, K. Berhane, F. Gilliland, D. Thomas, F. Lurmann, E. Avol, N. Kunzli, M. Jerrett, and J. Peters. 2007. Effect of Exposure to Traffic on Lung Development from 10 to 18 Years of Age: A Cohort Study. *Lancet* 369(9561): 571–7.

Geraghty, C.J. 1955. The Dismal Future of Smog Control. Presented to Inter-City Council of Santa Clara County, CA, January 13. (Speech copy courtesy of Anne Geraghty, California Air Resources Board).

Giovinazzo, C.T. 2003. California's Global Warming Bill: Will Fuel Economy Preemption Curb California's Air Pollution Leadership? *Ecol. Law Quarterly* 30: 893–954.

Godish, T. 1997. *Air Quality.* 3rd ed., Boca Raton, FL, and New York, NY: Lewis Publishers, Inc.

Goggin, M.L., A.O.M. Bowman, J.P. Lester, and L.J. O'Toole, Jr. 1990. *Implementation Theory and Practice: Toward a Third Generation.* HarperCollins Inc.

Gore, A. 1998. Speech prepared for the Senior Executive Service by Vice President Al Gore. Archive, National Partnership for Reinventing Government, Washington, DC, May 5.

Guckian, B.J. 2007. Personal Communication with Burford James Guckian. Austin, TX: Vehicle Inspection Bureau, Texas Department of Public Safety, April 25.

Gwilliam, K.M., M. Kojima, and T. Johnson. 2004. *Reducing Air Pollution from Urban Transport.* Washington, DC: World Bank.

Halperin, J. 1994. Springtime in Springfield: Campaign Rhetoric Shapes Agenda. Illinois Periodicals Online, Northern Illinois University Libraries, DeKalb, IL, January 26. http://www.lib.niu.edu/ipo/1994/ii940126.html. Accessed December 28, 2009.

Head, B.W. 2007. Community Engagement: Participation on Whose Terms? *Australian J. Poli. Sci.* 42(3): 441–54.

Hedglin, P. 2006. OBD Code Clearing: I/M Impact and Potential Solutions. Presented at the Colorado State University. *22nd Annual Clean Air Conference.* Breckenridge, CO, September 24–27.

Hertsgaard, M. 2002. California Green Light. *Nation* (August 19–26), 7.

Hileman, B. 1999. EPA's Enforcement Success Illusive. *Chem. & Eng. News* 77(37): 18–21.

Hill, M., and P. Hupe. 2002. *Implementing Public Policy: Governance in Theory and Practice.* Edited by I. Holliday, London, England: Sage Publications.

Hochgreb, S. 1998. Combustion-related emissions in SI engines. In *Handbook of Air Pollution from Internal Combustion Engines, Pollutant Formation and Control.* Edited by E. Sher, San Diego, CA: Academic Press, Chestnut Hill, MA, and London, England, pp. 118–70.

Holmes, K.J., D. Allen, and M. Russell. 2007. The Roles of State and Federal Mobile-source Emissions Standards. *Env. Sci. Technol.* 41(9): 3040–5.

Horn, J. 1995. Testimony of Jim Horn, Texas State Representative, to the Subcommittee on Oversight and Investigations of the Committee on Commerce, U.S. House of Representatives, March 24.

Houck, O. 2003. Tales from a Troubled Marriage: Science and Law in Environmental Policy. *Science* 302: 1926–9.

Howekamp, D. 1993. U.S. EPA's Comments on the I/M Review Committee's Draft Fourth Report to the Legislature. Letter from David P. Howekamp, U.S. Environmental Protection Agency, to Richard Sommerville, I/M Review Committee, January 26.

Hurley, D.R. 1992. Memorandum and Order. CV-92-1494. United States District Court, Eastern District of New York. Natural Resources Defense Council, Inc., Plaintiff, City of New York and State of New York, Intervenor-Plaintiffs, against United States Environmental Protection Agency; and William K. Reilly, as administrator of the United States Environmental Protection Agency, defendants. Brooklyn, New York, July 1.

Ingalls, M.N., L.R. Smith, and R.E. Kirksey. 1989. Measurement of On-road Vehicle Emission Factors in the California South Coast Air Basin. Vol. I. Regulated emissions. Prepared for the Coordinating Research Council, Atlanta, GA, by the Southwest Research Institute. No. SwRI-1604.

Inside EPA. 1994a. Marine I/M commotion. *Mobile Source Report,* September, 9.

———. 1994b. U.S. Senate Bill Seeks to Delay EPA Enhanced I/M Program. *Mobile Source Report,* December 2.

Irvin, R.A., and J. Stansbury. 2004. Citizen Participation in Decision Making: Is it Worth the Effort? *Public Admin. Rev.* 64(1): 55–65.

Jasanoff, S. 1990. *The Fifth Branch: Science Advisers as Policymakers.* Cambridge, MA, and London, England: Harvard University Press.

———. 1992. What Judges Should Know about the Sociology of Science. *Jurimetrics J. Law, Sci. & Technol.* 32 Spring: 345–59.

Jerrett, M., R.T. Burnett, C.A. Pope, K. Ito, G. Thurston, D. Krewski, Y. Shi, E. Calle, and M. Thun. 2009. Long-term Ozone Exposure and Mortality. *New Engl. J. Med.* 360(11): 1085–95.

Jordan, D.L. 1993. Newspaper Effects on Policy Preferences. *Pub. Opin. Quarterly* 57: 191–204.

Keller, K.M. 1995. State of California, Department of Consumer Affairs, Bureau of Automotive Repair, Sacramento, CA. Letter to Margo Oge. Washington, DC: U.S. Environmental Protection Agency, March 30.

Kenny, M.P. 2000. Letter from Michael P. Kenny, Executive Officer, California Air Resources Board, to Felicia Marcus. Washington, DC: U.S. Environmental Protection Agency, August 17.

Kettl, D.F. 1998. Reinventing Government: A Fifth-year Report card. Brookings Institution, Center for Public Management, Washington, DC: September.

Kingdon, J.W. 1984. *Agendas, Alternatives, and Public Policies.* Boston, MA, and Toronto, Canada: Little, Brown and Company.

Klausmeier, R. 2002. Literature and Best Practices Scan: Vehicle Inspection and Maintenance (I/M) programs. Final report prepared for the Wisconsin Department of Transportation, Bureau of Vehicle Services, Bureau of Environment, by de la Torre Klausmeier Consulting, Inc., Austin, TX. Project No. 0092-02-09, June.

Klausmeier, R., and P. McClintock. 2003. Virginia Remote Sensing Device Study. Final report prepared for Virginia Department of Environmental Quality, February. http://www.deq.state.va.us/air/pdf/air/rsdreport.pdf. Accessed December 28, 2009.

Klausmeier, R., S. Kishan, R. Baker, A. Burnette, and J. McFarland. 1995. Evaluation of the California Pilot Inspection/Maintenance (I/M) Program. Draft final report prepared for the California Bureau of Automotive Repair, Sacramento, CA, by de la Torre Klausmeier Consulting, Inc., Austin, TX, and Radian Corporation, March.

Klausmeier, R., S. Kishan, A. Burnette, and M. Weatherby. 2000. Smog Check Station Performance Analysis based on Roadside Test Results. Technical note prepared for the California Bureau of Automotive Repair, Engineering and Research Branch, Sacramento, CA, by de la Torre Klausmeier Consulting, Inc., and Eastern Research Group, Austin, TX, June 27.

Klausmeier, R., R. Oommen, J. Hauser, G. Manne, A. Burnette, and E. Gardetto. 2006. Colorado Automobile Inspection and Readjustment (AIR) Program. Colorado Department of Public Health and Environment performance audit prepared for the Office of the Colorado State Auditor, Denver, CO, by de la Torre Klausmeier Consulting, Inc., Eastern Research Group, and Sierra Research, November.

Kopp, Q.L. 1993. California state senator. Sacramento, CA. Letter to David P. Howekamp, Air Toxics Division, U.S. Environmental Protection Agency, Region IX, San Francisco, CA, April 12.

———. 1997. California State Senate Bill No. 42, Chapter 801: Air Pollution: Vehicles: Inspection and Maintenance, October 9.

Kraft, M.E., and N.J. Vig. 2003. Environmental policy from the 1970s to the twenty-first century. In *Environmental Policy: New Directions for the Twenty-first Century*. Edited by N.J. Vig and M.E. Kraft, Washington, DC: CQ Press pp. 1–32.

Krosnick, J.A., A.L. Holbrook, L. Lowe, and P.S. Visser. 2006. The Origins and Consequences of Democratic Citizens' Policy Agendas: A Study of Popular Concern about Global Warming. *Climatic Change*. 77: 7–43.

Kucharski, W.A. 1993. Enhanced I&M Program. Memorandum from W.A. Kucharski, State of Louisiana Department of Environmental Quality, to Gus Von Bodungen, State of Louisiana Department of Environmental Quality, November 24.

Lawson, D.R. 1993. "Passing the Test": Human behavior and California's Smog Check Program. *J. Air & Waste Manag. Assoc.* 43: 1567–75.

———. 1995. The costs of "M" in I/M: Reflections on Inspection/Maintenance Programs. *J. Air & Waste Manag. Assoc.* 45: 465–76.

———. 2005. Results from DOE's Gasoline/Diesel PM Split study and EPA's High Mileage OBD Study. Presented to the California I/M Review Committee, November 22.

Lawson, D.R., and E. Gardetto. 2006. Letter to the Editor. *J. Air & Waste Manag. Assoc.* 56: 242–3.

Lawson, D.R., P.J. Groblicki, D.H. Stedman, G.A. Bishop, and P.L. Guenther. 1990. Emissions from In-use Motor Vehicles in Los Angeles: A Pilot Study of Remote Sensing and the Inspection and Maintenance Program. *J. Air & Waste Manag. Assoc.* 40: 1096–105.

Lawson, D.R., P.A. Walsh, and P. Switzer. 1995. Effectiveness of U.S. Motor Vehicle Inspection/Maintenance Programs, 1985–1992. Final report. Prepared for the California I/M Review Committee, November 2.

Lawson, D.R., S. Diaz, E.M. Fujita, S.L. Wardenburg, R.E. Keislar, Z. Lu, and D.E. Schorran. 1996. Program for the Use of Remote Sensing Devices to Detect High-emitting Vehicles. Final report. Prepared for South Coast Air Quality Management District, Diamond Bar, CA, April 16.

Layzer, J.A. 2002. *The Environmental Case: Translating Values into Policy*. Washington, DC: CQ Press.

Liberatore, A. 2001. White Paper on Governance, Work Area 1, Broadening and Enriching the Public Debate on European Matters. Report of the Working Group "Democratising Expertise and Establishing Scientific Reference Systems" (Group 1b), July.

Lindblom, C.E. 1959. The Science of "Muddling Through." *Public Administration Review* 19(2, Spring): 79–88. American Society for Public Administration.

*Los Angeles Times*. 1992. 32 Arrested in Sting Aimed at Fraudulent Smog Checks, December 18.

———. 1993a. D.A. Sues Over Alleged Smog Violations, May 12.

———. 1993b. The Green Machine is Asleep at the Switch, November 1.

———. 1993c. Plan to Revamp State Smog Check Endorsed by EPA, August 27.

———. 1994a. EPA begins Steps to Penalize State Over Smog Plans, January 8.

———. 1994b. EPA rescinds Order, won't Cut off State Road Funds, January 25.

———. 1994c. Other States Check out California Smog Test Deal, March 24.

———. 2010. Smog in L.A., then and now. http://www.latimes.com/news/printedition/highway1/la-hy-smog-pg,0,4220006.photogallery. Accessed May 15, 2010.

Lowi, T.J. 1976. *American Government: Incomplete Conquest.* Hinsdale, IL: Dryden Press.

Lutsey, N., and D. Sperling. 2008. America's Bottom-up Climate Change Mitigation Policy. *Energy Policy* 36: 673–85.

Manza, J., and F.L. Cook. 2002. A Democratic Polity? Three Views of Policy Responsiveness to Public Opinion in the United States. *American Poli. Res.* 30: 630–67.

*Marin Independent Journal.* 1997. High-tech Sensor Sniffs Out Over-polluting Vehicles, February 20.

Matland, R.E. 1995. Synthesizing the Implementation Literature: The Ambiguity-conflict Model of Policy Implementation. *J. Public Admin. Research & Theory*, Part V (April): 145–74. Reprinted in *Public Administration: Concepts and Cases.* 7th ed., January 2000. Edited by R.J. Stillman, Houghton Mifflin Co., Boston, MA, pp. 407–425.

Mazmanian, D.A., and P.A. Sabatier. 1989. *Implementation and Public Policy.* Lanham, MD: University Press of America, Inc.

McClintock, P.M. 2006. Virginia Remote Sensing Device. Draft 2005 Annual Report prepared for the Virginia Department of Environmental Quality, Richmond, VA, by ESP, Tucson, AZ, March.

McClintock, P. 2007. Remote Sensing: Another Point of View. Presented at the California I/M Review Committee Meeting by Applied Analysis, Tiburon, CA, June 26.

McClintock, P., and H. Vescio. 2007. RSD Technical Feasibility. Presented at the Smog Check Technology Forum & Round Table, South Coast Air Quality Management District, Diamond Bar, CA, by Applied Analysis, Tiburon, CA, and ESP, East Granby, CT, March 21. http://www.aqmd.gov/tao/ConferencesWorkshops/SmogCheckForum/5_McClintock_Slides.pdf. Accessed December 28, 2009.

Melnick, R.S. 1983. *Regulation and the Courts: The Case of the Clean Air Act.* Washington, DC: Brookings Institution.

Migden, C. 1997. California State Assembly Bill No. 208, Chapter 802: Vehicles: Inspection and Maintenance: High Polluter Repair or Removal, October 9.

Milton, B.E. 1998. Control Technologies in Spark-ignition Engines. In *Handbook of Air Pollution from Internal Combustion Engines, Pollutant Formation and Control.* Edited by E. Sher. San Diego, CA, Chestnut Hill, MA, and London, England: Academic Press, pp. 189–258.

Ministry of Environment and Forests. 2005. Central Pollution Control Board, 2004–2005 Annual Report. Government of India, http://www.cpcb.nic.in/annual_archive.php?pno=1. Accessed October 30, 2009.

M.J. Bradley & Associates Inc. 2002. Air Toxics Benefits From Vehicle Inspection and Maintenance Programs in Select U.S. Cities. Report prepared for the American Lung Association, Washington, DC, October.

MO DNR (Missouri Department of Natural Resources). 2005. Emissions Inspection and Maintenance Summit. Draft white paper, October 27.

Morrow, S., and K. Runkle. 2005. April 2004. Evaluation of the California's Enhanced Vehicle Inspection and Maintenance (Smog Check) Program. Final report (revised) prepared for the California State Legislature, Addendum completion date, September.

Naisbitt, J. 1982. *Megatrends: Ten New Directions Transforming Our Lives.* New York: Warner Books.

NAS (National Academy of Sciences). 2000. *Strengthening Science at the U.S. Environmental Protection Agency: Research Management and Peer Review Practices.* National Academy of Sciences, Washington, DC: Contract No. 68-W4-0044.

NPRG (National Partnership for Reinventing Government). 1993. Remarks by President Clinton Announcing the Initiative to Streamline Government, March 3, Archive, University of North Texas Libraries.

Nelkin, D. 1987. *Selling Science: How the Press Covers Science and Technology.* New York: W.H. Freeman and Company, pp. 70–84.

Nemery, B., P.H. Hoet, and A. Nemmar. 2001. The Meuse Valley fog of 1930: An Air Pollution Disaster. *Lancet* 357(March 3): 704–8.

*New York Times.* 1992. The Citizen as Customer, March 8.

———. 1993. States and Federal Agencies Lag in Meeting Clean Air law, November 16.

———. 1994. E.P.A. Critics Get Boost in Congress, February 7.

Nichols, M.D. 1993. Statement of Mary Nichols. U.S. Environmental Protection Agency, issued by Denise Graveline, EPA Region 9, November 24.

———. 1995a. Statement of Mary D. Nichols. U.S. Environmental Protection Agency, before the Subcommittee on Oversight and Investigations of the Committee on Commerce, U.S. House of Representatives, March 24.

———. 1995b. Letter from Mary D. Nichols. U.S. Environmental Protection Agency, to George W. Bush, governor of Texas, April 11.

NRC (National Research Council). 1992. *Rethinking the Ozone Problem in Urban and Regional Air Pollution.* Washington, DC: National Academies Press.

———. 2000. *Modeling Mobile Source Emissions.* Washington, DC: National Academies Press.

———. 2001. *Evaluating Vehicle Emissions Inspection and Maintenance Programs.* Washington, DC: National Academies Press.

———. 2004. *Air Quality Management in the United States.* Washington, DC: National Academies Press.

NRC (National Research Council) Committee on Risk Characterization, 1996. Deliberation. In *Understanding Risk: Informing Decisions in a Democratic Society.* Edited by P.C. Stern, H.V. Fineberg. Washington, DC: National Academies Press, pp. 73–96.

Nye, J.S. 2008. *The Powers to Lead.* New York: Oxford University Press.

Olin, R. 2007. Personal communication with Richard Olin, Virginia Department of Environmental Quality, Mobile Source Program, Richmond, VA, May 4.

Osborne, D., and T. Gaebler. 1992. *Reinventing Government: How the Entrepreneurial Spirit is Transforming the Public Sector. From Schoolhouse to Statehouse, City Hall to the Pentagon.* Reading, MA: Addison-Wesley Publishing Company, Inc.

PA DEP (Pennsylvania Department of Environmental Protection). 1995. Gov. Ridge Signs Bill to Implement Envirotest Settlement. Press Release, Pennsylvania Department of Environmental Protection, December 15. http://www.dep.state.pa.us/dep/deputate/polycomm/pressrel/95/Govsigns.htm. Accessed December 28, 2009.

Peace, S. 1994. Legislative Debate on California State Senate Bill 521, March 17.

Pidgeon, W.M., D.J. Sampson, P.H. Burbage, L.C. Landman, W.B. Clemmens, E. Herzog, D.J. Brzezinski, and D. Sosnowski. 1993. Evaluation of a Four-mode Steady-state Test with Acceleration Simulation Modes as an Alternative Inspection and Maintenance Test for Enhanced I/M programs. Research Triangle Park, NC: EPA-AA-AQAB-93-01 (NTIS PB94-129905). U.S. Environmental Protection Agency. May.

Pierson, W.R. 1996. Motor Vehicle Inspection and Maintenance Programs—How Effective are they? *Atmospheric Env.* 30(21): i–iii.

Pierson, W.R., A.W. Gertler, and R.L. Bradow. 1990. Comparison of the SCAQS Tunnel Study with other On-road Vehicle Emission Data. *J. Air & Waste Manag. Assoc.* 40(11): 1495–504.

Pierson, W.R., D.E. Schorran, E.M. Fujita, J.C. Sagebiel, D.R. Lawson, and R.L. Tanner. 1999. Assessment of Nontailpipe Hydrocarbon Emissions from Motor Vehicles. *J. Air & Waste Manag. Assoc.* 49: 498–519.

Pinch, T. 2000. The Golem: Uncertainty and Communicating Science. *Sci. & Eng. Ethics* 6(4): 511–23.

Pitchford, M., and B. Johnson. 1993. Empirical Model of Vehicle Emissions. *Env. Sci. Technol.* 27: 741–8.

Pollack, A.K., T.E. Stoeckenius, T. Wenzel, J. Schwartz, and V. McConnell. 2003. Performance Audit of the Colorado Automobile Inspection and Readjustment (AIR) Program. Final report prepared for the Office of the State of Colorado Auditor, Denver, CO, June 30.

Portney, P.R. 1990. Overall Assessment and Future Directions. In *Public Policies for Environmental Protection.* Edited by P.R. Portney. Resources for the Future, Washington, DC, pp. 275–89.

Powell, M.R. 1999. *Science at EPA: Information in the Regulatory Process.* Resources for the Future, Washington, DC.

Presley, R. 1993. California State Senate Bill No. 119: Motor Vehicle Inspection Program. Introduced by California senator Robert Presley, Sacramento, CA, January.

Rabe, B.G. 2003. Power to the States: The Promise and Pitfalls of Decentralization. In *Environmental Policy: New Directions for the Twenty-first Century.* Edited by N.J. Vig, M.E. Kraft, Washington, DC: CQ Press, pp. 33–56.

———. 2007. Environmental Policy and the Bush era: The Collision between the Administrative Presidency and State Experimentation. *Publius* 37(3): 413–31.

RAQC (Regional Air Quality Council). 2000. Carbon Monoxide Maintenance Plan Approved. AIR Exchange Newsletter. Regional Air Quality Council, Denver, CO: Winter–Spring.

Reitze, Jr. A.W. 1979. Controlling Automotive Air Pollution through Inspection and Maintenance Programs. *George Washington Law Rev.* 47(4): 705–39.

———. 1996. Federalism and the Inspection and Maintenance Program under the Clean Air Act. *Pacific Law Journal* 27(4): 1461–520.

Roberts, N. 2004. Public Deliberation in an Age of Direct Citizen Participation. *American Rev. Public Admin.* 34(4): 315–53.

*Rolling Stone.* 1993. The Sinkable Carol Browner, New EPA Chief Finds Herself in over Her Head, p. 45, October 28.

Rosenbaum, W.A. 2003. Still Reforming after all these Years: George W. Bush's "New Era" at the EPA. In *Environmental Policy: New Directions for the Twenty-first Century.* Edited by N.J. Vig, and M.E. Kraft. Washington, DC: CQ Press, pp. 175–99.

Russell, N.R. 1993. Senate Bill No. 1195: Vehicles: Inspection and Maintenance. Introduced by California Senator Russell and Principal Coauthor, California Senator Boatwright, Sacramento, CA, March.

Russell, N.R., D. Boatwright, and R. Katz. 1994. California State Senate Bill No. 629, Chapter 1. Vehicles: Inspection and Maintenance, January 28.

Sabatier, P.A. 1986. Top-down and Bottom-up Approaches to Implementation Research: A Critical Analysis and Suggested Synthesis. *J. Public Policy* 6: 21–48.

———. 1988. An Advocacy Coalition Framework of Policy Change and the Role of Policy-oriented Learning Therein. *Policy Sci.* 21: 129–68.

Sabatier, P.A., H.C. Jenkins-Smith, H.B. Mawhinney, A.E. Brown, J. Stewart, Jr., J.F. Munro, R.P. Barke, G.K. Clair St., and A.M. Brasher. 1993. In *Policy Change and Learning: An Advocacy Coalition Approach.* Edited by P.A. Sabatier, and H.C. Jenkins-Smith, Boulder, CO: Westview Press, Inc.

*Sacramento Bee.* 1993. Wilson won't Support Pact on Smog Checks, August 31.

Saito, D. 2007. Western Riverside County Clean Cities. Session 1, Track 2: Regulations, Policies and Upcoming Programs. Presented at the 8th Annual Advancing the Choice Expo: An Alternative to Petroleum Based Fuels, Riverside, CA, by the South Coast Air Quality Management District, Diamond Bar, CA, February 8.

*San Diego Union-Tribune.* 1993a. Get Smog Plan or Lose Funds, State Told, April 16.

———. 1993b. Smog-Check Standoff with EPA Grows Testy, August 14.

———. 1994. Smog Check Blowout, EPA's Pressure Tactics Backfire Badly, January 14.

*San Francisco Chronicle.* 1993. Skirmish over Change in Smog Law, May 24.

———. 1994a. EPA Turns up the Heat on State in Smog-test War, January 8.

———. 1994b. Compromise on Smog Checks, only 15% of Cars will be Required to Use Inspection-only Stations, March 10.

———. 1996. California Chokes on New Smog Law, August 20.

*San Jose Mercury News.* 1993a. EPA's Smog Stand Lacks Factual Basis, Some Experts Say, September 16.

———. 1993b. Smog Check Rebellion Could set an Example, December 26.

———. 1994. Compromise Near on New Smog Check, February 14.

Santa Clara University. 2008. Most cited U.S. Supreme Court Cases. http://law.scu.edu/library/most-cited-us-supreme-court-cases.cfm. Accessed December 28, 2009.

SCAQMD (South Coast Air Quality Management District). 1997. The Southland's War on Smog, *Diamond Bar.* May.

———. 2005. Board Meeting Agenda Item No. 6: Execute Sole Source Contract for Voluntary Vehicle Emission Testing and Repair Assistance Rrogram, Issue RFP for High Emitting Light- and Medium-duty Vehicle Identification Program, Issue Voluntary High Emitting Vehicle Scrapping Program Announcement, and Establish Voluntary Vehicle Scrapping Incentive Program and Contingency Fund, September 9.

———. 2006. Board Meeting Agenda Item No. 5: Execute Contract and Agreement to Implement High Emitter Identification Program and Voluntary High Emitting Vehicle Scrapping Program, February 3.

———. 2008. Multiple Air Toxics Exposure Study in the South Coast Air Basin, MATES-III. Final report. Diamond Bar, CA, September.

Schifter, I., L. Diaz, J. Duran, E. Guzman, O. Chavez, and E. Lopez-Salinas. 2003. Remote Sensing Study of Emissions from Motor Vehicles in the Metropolitan Area of Mexico City. *Env. Sci. Technol.* 37(2): 395–401.

Schmitt, R.R. 1998. A Brief Overview of the Northridge Earthquake and its Transportation Impacts. *J. Transp. Stats.* 1(2): v–vi.

Sclove, R.E. 1995. *Democracy and Technology.* Chapter 3. New York: Guilford Press, pp. 25–57.

Sierra Research. 2006. Status Report: Smog Check Program Evaluation Project, October 24.

Singer, B.C., and T.P. Wenzel. 2003. Estimated Emission Reductions from California's Enhanced Smog Check Program. *Env. Sci. Technol.* 37(11): 2588–95.

Sjodin, A., and M. Lenner. 1995. On-road Measurements of Single Vehicle Pollutant Emissions, Speed and Acceleration for Large Fleets of Vehicles in Different Traffic Environments. *Sci. Total Env.* 169: 157–65.

South Coast Air Quality Management District. 1997. The Southland's War on Smog. Diamond Bar. May.

Spencer, L. 1992. Not Invented Here. *Forbes* (October 12): 44–5.

Sperling, D., and D. Gordon. 2009. *Two Billion Cars.* Oxford: Oxford University Press.

SRI (Stanford Research Institute). 1949. The Smog Problem in Los Angeles County. Second Interim Report by Stanford Research Institute on Studies to Determine the Nature and Sources of the Smog. Committee on Smoke and Fumes, Western Oil and Gas Association, August.

State of Missouri. 2007. Gateway Vehicle Inspection Program. Request for Proposal, REQ NO. NR 780 34057000003. State of Missouri, Jefferson City, February.

Stedman, D.H. 2006. Letter to the Editor. *J. Air & Waste Manag. Assoc.* 56: 242.

Stedman, D.H., and G.A. Bishop. 2004. A History of On-road Emissions and Emissions Deterioration. Presented at the 14th Coordinating Research Council On-Road Vehicle Emissions Workshop, San Diego, CA, March 31.

Stedman, D.H., G. Bishop, J.E. Peterson, and P.L. Guenther. 1991. On-road CO Remote Sensing in the Los Angeles Basin. Final report prepared for the California Air Resources Board, Sacramento, CA, Contract No. A932-189, August.

Stoker, R.P. 1991. *Reluctant Partners: Implementing Federal Policy.* Pittsburgh, PA: University of Pittsburgh Press.

Strock, J.M., and J.C. Kozberg. 1995. Letter from James M. Strock, California Secretary for Environmental Protection, and Joanne C. Kozberg, California Secretary for State and Consumer Services, to Carol M. Browner, administrator. U.S. Environmental Protection Agency, Washington, DC: March 13.

Strock, J., and A. Poat. 1992. Letter from James M. Strock, California Secretary for Environmental Protection, and Andrew Poat, California Undersecretary, State and Consumer Services Agency, to U.S. Environmental Protection Agency, Washington, DC. Air Docket No. A-91-75, Notice of Proposed Rulemaking, Subpart S: Vehicle Inspection and Maintenance Requirements for State Implementation Plans, August 26.

TCEQ (Texas Commission on Environmental Quality). 2001. SIP Revision: Vehicle Inspection/Maintenance. October 24. http://www.tceq.state.tx.us/implementation/air/sip/oct2001im.html. Accessed December 28, 2009.

Thomas, L.M. 1987. Next Steps in the Battle Against Smog. *EPA Journal* 13(8): 2–4.

Tierney, G. 2007. Transitioning I/M: Options for I/M in the OBD Dominated Fleet. Presented at the 23rd Annual Clean Air Conference, National Center for Vehicle Emissions Control and Safety, Colorado State University, Breckenridge, CO, September 26.

TRB (Transportation Research Board). 2002. *The Congestion Mitigation and Air Quality Improvement Program: Assessing 10 Years of Experience.* Special report 264. Washington, DC: National Academies Press.

U.S. AID (U.S. Agency for International Development). 2004. Vehicle Inspection and Maintenance Programs: International Experience and Best Practices, October. http://pdf.usaid.gov/pdf_docs/PNADB317.pdf. Accessed May 18, 2010.

U.S. EPA (U.S. Environmental Protection Agency). 1990. Approval and Promulgation of Implementation Plans for Ozone and Carbon Monoxide, California (South Coast Air Basin). Proposed Rule. *Federal Register*, 55(172): 36458–36576. 40 CFR Parts 51 and 52, September 5.

———. 1992a. Air Pollution from Motor Vehicles Studied at Fort McHenry Tunnel. *Environmental News*, press release. U.S. Environmental Protection Agency, Office of Community Outreach, Research Triangle Park, NC. June.

———. 1992b. EPA Toughens Emission Test for Cars and Trucks. *Environmental News for Release Thursday, November 5, 1992*. Washington, DC.

———. 1992c. *Highway Vehicle Emission Estimates*. U.S. Environmental Protection Agency, Office of Mobile Sources, Ann Arbor, MI. June.

———. 1992d. I/M Costs, Benefits, and Impacts. Washington, DC, November. http://www.epa.gov/otaq/regs/im/im-tsd.pdf. Accessed December 28, 2009.

———. 1992e. Inspection/maintenance Program Requirements. Final Rule. *Federal Register* 57(215): 52949–3014. 40 CFR Part 51, November 5.

———. 1993a. Control of Air Pollution from New Motor Vehicles and New Motor Vehicle Engines: Evaporative Emission Regulations for Gasoline- and Methanol-fueled Light-duty Vehicles, Light-duty Trucks and Heavy-duty Vehicles. Final Rule. *Federal Register*. 58: 16002 March.

———. 1993b. Quantitative Assessments of Test-only and Test-and-repair I/M Programs. Washington, DC. EPA-AA-EPSD-I/M-93-1, November.

———. 1993c. Some California I/M Options. Staff papers. Office of Mobile Sources, December.

———. 1994a. Memorandum of Agreement between the California Environmental Protection Agency and the United States Environmental Protection Agency. Signed by James Strock, California Environmental Protection Agency, and Carol Browner, U.S. Environmental Protection Agency, March.

———. 1994b. U.S. EPA Oral History Interview 3: Alvin L. Alm. Public Information Center, Washington, DC. EPA 202-K-94-005, January.

———. 1994c. U.S. EPA Proposes Sanctions against three states under Clean Air Act. Press release. Washington, DC, January 7.

———. 1995a. Air Division Director Briefing Materials, Maine I&M. Undated document prepared by U.S. Environmental Protection Agency, Office of Mobile Sources, Ann Arbor, MI, for distribution to EPA Regional Air Division directors.

———. 1995b. Final Technical Report on Aggressive Driving Behavior for the Revised Federal Test Procedure Notice of Proposed Rulemaking. Office of Mobile Sources. January. http://www.epa.gov/oms/regs/ld-hwy/ftp-rev/ftp-us06.pdf. Accessed December 28, 2009.

———. 1995c. *I/M Briefing Book*. EPA-AA-EPSD-I/M-94-1226. U.S. Government Printing Office, Washington, DC.

———. 1995d. Inspection/Maintenance Flexibility Amendments. Final Rule. *Federal Register* 60(180): 48029–37. 40 CFR Part 51, Septemper 18.

———. 1996a. Inspection/Maintenance Flexibility Amendments (Ozone Transport Region). Final Rule. *Federal Register* 61(144): 39032–7. 40 CFR Part 51, July 25.

———. 1996b. Motor Vehicle Emissions Federal Test Procedure Revisions. *Federal Register* 61(205): 54852–906. 40 CFR Part 86, October 22.

———. 1999. Major Elements of Operating I/M Programs. Table prepared by the U.S. Environmental Protection Agency, Office of Mobile Sources, Research Triangle Park, NC, EPA420-B-99-008, December.

———. 2000a. Additional Flexibility Amendments to Vehicle Inspection Maintenance Program Requirements. Amendment to the Final Rule. *Federal Register.* 65(142): 45526–35. 40 CFR Part 51, July 24.

———. 2000b. Analyses of the OBDII Data Collected from the Wisconsin I/M Lanes. Technical report prepared by Ted Trimble, Transportation and Regional Programs Division, Office of Transportation and Air Quality, Research Triangle Park, NC, EPA420-R-00-014. August.

———. 2000c. Control of Air Pollution from New Motor Vehicles: Tier 2 Motor Vehicle Emissions Standards and Gasoline Sulfur Control Requirements. Final Rule. *Federal Register* 65(28): 6698–822. 40 CFR Parts 80, 85, and 86, February 10.

———. 2001a. Amendments to Vehicle Inspection Maintenance Program Requirements Incorporating the Onboard Diagnostic Check. Final Rule. *Federal Register* 66(66): 18156–79. 40 CFR Parts 51 and 85, April 5.

———. 2001b. Clean Air Act. http://www.epa.gov/air/caa//. Accessed April 26, 2008.

———. 2002. Dynamometer Driver's Aid and Dynamometer Driving Schedules, August. http://www.epa.gov/nvfel/testing/dynamometer.htm. Accessed April 26, 2008.

———. 2003a. EPA's Budget and Workforce, 1970–2003. Budget Division, November. http://www.epa.gov/history/org/resources/budget.htm. Accessed April 26, 2008.

———. 2003b. Major Elements of Operating I/M Programs. Table prepared by the U.S. Environmental Protection Agency, Office of Transportation and Air Quality, Research Triangle Park, NC, EPA420-B-03-012. March.

———. 2003c. 2003–2008 EPA Strategic Plan: Direction for the Future. Office of the Chief Financial Officer, September. http://www.epa.gov/ocfo/plan/2003sp.pdf. Accessed April 26, 2008.

———. 2004a. Description and History of the MOBILE Highway Vehicle Emission Factor Model. Office of Transportation and Air Quality, Research Triangle Park, NC, February. http://www.epa.gov/otaq/models/mob_hist.txt. Accessed April 26, 2008.

———. 2004b. FY 2003 Annual Report. Supplemental Information for Sustained Progress in Addressing Management Issues. Office of Planning, Analysis, and Accountability, Office of the Chief Financial Officer, January.

———. 2007. The Cost-Effectiveness of Heavy-duty Diesel Retrofits and Other Mobile Source Emission Reduction Projects and Programs. Office of Transportation and Air Quality, Research Triangle Park, NC, EPA420-B-07-006. May.

———. 2008. IM OBD Vehicles Readiness Exception List. In *Office of Transportation and Air Quality*. EPA420-F-08-009, January.

———. 2009a. 8-hour Ozone Area Summary. Green book, Research Triangle Park, NC. December. http://www.epa.gov/oar/oaqps/greenbk/gnsum.html. Accessed August 29, 2009.

———. 2009b. National Emission Inventory (NEI) Air Pollutant Emissions Trends Data. http://www.epa.gov/ttn/chief/trends/. Accessed August 30, 2009.

———. 2009c. Particulate Matter ($PM_{2.5}$) Nonattainment Area Summary. Green book, Research Triangle Park, NC. December. http://www.epa.gov/oar/oaqps/greenbk/qnsum.html. Accessed August 29, 2009.

———. 2010. Inventory of U.S. Greenhouse Gas Emissions and Sinks. USEPA #430-R-10-006, April. http://epa.gov/climatechange/emissions/downloads10/US-GHG-Inventory-2010_Report.pdf. Accessed May 18, 2010.

U.S. GAO (U.S. General Accounting Office). 1997. Environmental Protection: Challenges Facing EPA's Efforts to Reinvent Environmental Regulation. Washington, DC. GAO/RCED-97-155, July.

———. 2000. Air Pollution: Status of Implementation and Issues of the Clean Air Act Amendments of 1990. Washington, DC. GAO/RCED-00-72, April.

———. 2001. Environmental Protection Agency: Status of Achieving Key Outcomes and Addressing Major Management Challenges. Washington, DC. GAO-01-774, June.

USGS (U.S. Geological Survey). 1996. USGS response to an Urban Earthquake: Northridge '94. Prepared for the Federal Emergency Management Agency, Open-file Report 96-263.

VA DEQ (Virginia Department of Environmental Quality). 2006a. I&M Program Evaluation. Virginia Department of Environmental Quality, Richmond, VA. http://www.deq.state.va.us/mobile/mobeval.html. Accessed December 28, 2009.

———. 2006b. On-road Emissions Testing Program Status. A Report to the Virginia General Assembly Prepared in Response to 2006 House Joint Resolution 208. December. http://www.deq.virginia.gov/regulations/pdf/2007.on.road.emissions.testing.program.status.report.pdf. Accessed December 28, 2009.

———. 2007. On-road Emissions Program: Frequently Asked Questions. Virginia Department of Environmental Quality, Woodbridge, VA. http://www.deq.state.va.us/mobile/orefaq.html. Accessed December 28, 2009.

*Wall Street Journal*. 1993. California, EPA Close to Accord on Auto Tests, November 26.

———. 1994. New Flexibility EPA on Emissions puts Texas Officials in a Costly Bind, December 14.

———. 1995. Not in My Garage: Clean Air Act Triggers Backlash as its Focus Shifts to Driving Habits, January 25.

Wang, T. 2006. High Levels of Particulate Pollution in Chinese Megacities. *Env. Sci. Technol.* 40(15): 4532–3.

*Washington Post.* 1994a. EPA may let Va. Emissions Test System Continue, March 14.

———. 1994b. EPA Yields to Governors on Auto Emissions Tests, December 10.

Watson, J.G., E.M. Fujita, J.C. Chow, B. Zielinska, L.W. Richards, W. Neff, and D. Dietrich. 1998. Northern Front Range Air Quality Study. Final report prepared for Colorado State University, Cooperative Institute for Research in the Atmosphere, Fort Collins, CO. June.

Waxman, H.A. 1992. The Clean Air Act of 1990: An Overview of its History and Policy. In *Clean Air Law and Regulation.* Edited by T.A. Vanderver. Washington, DC: Bureau of National Affairs, pp. 20–42.

WBCSD (World Business Council for Sustainable Development). 2004. Mobility 2030: Meeting the Challenges to Sustainability: The Sustainable Mobility Project. http://www.wbcsd.org/web/publications/mobility/mobility-full.pdf. Accessed May 15, 2010.

Weisser, V. 2006. Letter to Shirley Horton, California assemblywoman, from the California Inspection and Maintenance Review Committee, Sacramento, CA, May 2.

Wenzel, T., and R. Sawyer. 1998. Analysis of a Remote Sensing Clean Screen Program in Arizona. Prepared for U.S. Department of Energy by Lawrence Berkeley National Laboratory, University of California, LBNL-41918. Berkeley, CA. October.

Wilson, P. 1994. Approval of SB 629 (Smog Check) by California Governor Pete Wilson. Letter to members of the California State Senate, January 27.

Wurzel, R.K. 2002. *Environmental Policy-making in Britain, Germany, and the European Union: The Europeanisation of Air and Water Pollution Control.* Manchester, England: Manchester University Press.

APPENDIX A

# ENHANCED I/M REQUIREMENTS IN THE 1990 CAAA

Following is the full text of the U.S. statutory requirements for enhanced inspection and maintenance programs, as originally included in the 1990 Clean Air Act Amendments (CAAA), Section 182(c)(3):

**c) Serious Areas**

Except as otherwise specified in paragraph (4), each State in which all or part of a Serious Area is located shall, with respect to the Serious Area (or portion thereof, to the extent specified in this subsection), make the submissions described under subsection (b) (relating to Moderate Areas), and shall also submit the revisions to the applicable implementation plan (including the plan items) described under this subsection . . .

**(3) Enhanced vehicle inspection and maintenance program**

**(A) Requirement for submission**

Within 2 years after the date of the enactment of the Clean Air Act Amendments of 1990, the State shall submit a revision to the applicable implementation plan to provide for an enhanced program to reduce hydrocarbon emissions and $NO_x$ emissions from in-use motor vehicles registered in each urbanized area (in the nonattainment area), as defined by the Bureau of the Census, with a 1980 population of 200,000 or more.

**(B) Effective date of state programs; guidance**

The State program required under subparagraph (A) shall take effect no later than 2 years from the date of the enactment of the Clean Air Act Amendments of 1990, and shall comply in all respects with guidance published in the Federal Register (and from time to time revised) by the Administrator for enhanced vehicle inspection and maintenance programs. Such guidance shall include-

(i) a performance standard achievable by a program combining emission testing, including on-road emission testing, with inspection to detect tampering with emission control devices and misfueling for all light-duty vehicles and all light-duty trucks subject to standards under section 202; and
(ii) program administration features necessary to reasonably assure that adequate management resources, tools, and practices are in place to attain and maintain the performance standard.

Compliance with the performance standard under clause (i) shall be determined using a method to be established by the Administrator.

**(C) State program**

The State program required under subparagraph (A) shall include, at a minimum, each of the following elements-

(i) Computerized emission analyzers, including on-road testing devices.
(ii) No waivers for vehicles and parts covered by the emission control performance warranty as provided for in section 207(b) unless a warranty remedy has been denied in writing, or for tampering-related repairs.
(iii) In view of the air quality purpose of the program, if, for any vehicle, waivers are permitted for emissions related repairs not covered by warranty, an expenditure to qualify for the waiver of an amount of $450 or more for such repairs (adjusted annually as determined by the Administrator on the basis of the Consumer Price Index in the same manner as provided in title V).

(iv) Enforcement through denial of vehicle registration (except for any program in operation before the date of the enactment of the Clean Air Act Amendments of 1990 whose enforcement mechanism is demonstrated to the Administrator to be more effective than the applicable vehicle registration program in assuring that noncomplying vehicles are not operated on public roads).

(v) Annual emission testing and necessary adjustment, repair, and maintenance, unless the State demonstrates to the satisfaction of the Administrator that a biennial inspection, in combination with other features of the program which exceed the requirements of this Act, will result in emission reductions which equal or exceed the reductions which can be obtained through such annual inspections.

(vi) Operation of the program on a centralized basis, unless the State demonstrates to the satisfaction of the Administrator that a decentralized program will be equally effective. An electronically connected testing system, a licensing system, or other measures (or any combination thereof) may be considered, in accordance with criteria established by the Administrator, as equally effective for such purposes.

(vii) Inspection of emission control diagnostic systems and the maintenance or repair of malfunctions or system deterioration identified by or affecting such diagnostics systems.

Each State shall biennially prepare a report to the Administrator which assesses the emission reductions achieved by the program required under this paragraph based on data collected during inspection and repair of vehicles. The methods used to assess the emission reductions shall be those established by the Administrator.

APPENDIX B

# AUTOMOTIVE AIR POLLUTION AND EMISSION STANDARDS

## WHY VEHICLES POLLUTE: A BRIEF I/M-RELATED PRIMER

Preferably, fossil-fueled vehicles would consume every molecule of fuel in the combustion process, and automobiles would produce only carbon dioxide ($CO_2$) and water ($H_2O$). Of course, $CO_2$ emissions are problematic for climate change reasons; however, in this discussion, $CO_2$ is ignored to focus on automotive pollutants that contribute to urban-scale pollution and were addressed by I/M programs as of the late 2000s. Motor vehicles emit hundreds of substances; the most important at the regional scale include hydrocarbons (HC), oxides of nitrogen ($NO_x$), carbon monoxide (CO), particulate matter (PM), and air toxics. Two factors in particular play an important role in explaining why automobiles emit pollutants other than $CO_2$ and $H_2O$: the composition of ambient air and the optimal air-to-fuel ratio to combust fuel.

Tropospheric (ground-level) air is composed of 78 percent nitrogen (N) and 21 percent oxygen (O). At high temperatures, such as those encountered in an internal combustion engine, nitrogen and oxygen combine to form $NO_x$, including nitric oxide (NO) and nitrogen dioxide ($NO_2$). Even if an automobile efficiently consumed fuel, the vehicle would be expected to emit $NO_x$ given the overwhelming presence of nitrogen in air and high combustion temperatures.

The ability of a vehicle to efficiently combust fuel and limit emissions is dependent on the engine and vehicle design, as well as its maintenance. In gasoline-powered vehicles, some of the fuel can evaporate and be lost prior to combustion. Older vehicles suffer from less efficient engines. Even late-model vehicles do not combust fuel efficiently when engine components fail or are improperly maintained. Conceptually, if some fraction of the hydrocarbon fuel remains unburned, then vehicle emissions will include HC in addition to $CO_2$, $H_2O$, and $NO_x$. In addition to HC emissions, incomplete fuel combustion also produces CO.

It is possible to estimate the optimal air-to-fuel ratio required to fully combust fuel by estimating the amount of air needed for complete fuel combustion. For example, complete combustion of gasoline in the presence of air can be represented in simplified form (ignoring $NO_x$ formation, for example) by Equation 1, using the formula for octane, $C_8H_{18}$, to represent the average molecular formula for all gasoline (for more details, see Godish 1997; Milton 1998):

$$2\,C_8H_{18} + 25\,O_2 + 93\,N_2 = 16\,CO_2 + 93\,N_2 + 18\,H_2O \qquad (1)$$

Based on molecular weights of 12, 1, 16, and 14 for C, H, O, and N, respectively, to combust 228 grams of gasoline (approximated here by 2 $C_8H_{18}$) requires 3,404 grams of air, or an air-to-fuel ratio of about 15:1. More precise estimates of this ratio are defined as between 14.5:1 and 14.7:1 and are detailed elsewhere (e.g., U.S. EPA 1995b; Hochgreb 1998).

The concepts of optimal fuel combustion and air-to-fuel ratios have significant implications for air pollution control. Gasoline-powered vehicles operating at or near optimum air-to-fuel ratios are likely to minimize HC and CO emissions, but maximize $NO_x$ emissions. A second implication is the difficulty of controlling emissions based solely on air-to-fuel considerations, as different driving conditions require different air-to-fuel mixtures. Gasoline-powered vehicles traditionally have provided maximum power output when running rich—for example, with air-to-fuel ratios in the range of 12.5:1 to 14:1. Thus, during periods when vehicles are accelerating, passing, climbing hills, operating under heavy loads, or otherwise in need of power, the air-to-fuel ratio is likely to

| *Model year* | *Federal* | | | | | *California* | | | | |
|---|---|---|---|---|---|---|---|---|---|---|
| | *HC* | *CO* | *NO$_x$* | *Evap*[a] | *PM* | *HC* | *CO* | *NO$_x$* | *Evap*[a] | *PM* |
| 1966 | | | | | | 6.30 | 51.0 | | | |
| 1968 | 6.3 | 51.0 | | | | | | | | |
| 1970 | 4.1 | 34.0 | | | | 4.1 | 34.0 | | 6 | |
| 1971 | | | | | | | | 4.0 | | |
| 1972 | 3.0 | 28.0 | | | | 2.9 | 34.0 | 3.0 | 2 | |
| 1973 | | | 3.0 | | | | | | | |
| 1974 | | | | | | | | 2.0 | | |
| 1975 | 1.5 | 15 | 3.1 | 2 | | 0.9 | 9.0 | | | |
| 1977 | | | 2.0 | | | 0.41 | | 1.5 | | |
| 1978 | | | | 6 | | | | | 6 | |
| 1980 | 0.41 | 7.0 | | | | 0.39 | | 1.0 | 2 | |
| 1981 | | 3.4 | 1.0 | 2 | | | 7.0 | 0.7 | | |
| 1983 | | | | | | | | 0.4 | | |
| 1984 | | | | | | | | | | 0.6 |
| 1985 | | | | | | | | | | 0.4 |
| 1986 | | | | | | | | | | 0.2 |
| 1988 | | | | | | | | | | |
| 1989 | | | | | | | | | | 0.08 |
| 1993 | | | | | | 0.25[b] **(0.31)** | 3.4[b] **(4.2)** | | | |

| | | | | | | | | | | |
|---|---|---|---|---|---|---|---|---|---|---|
| | Tier 1 U.S. and LEV I California standards | | | | | | | | | |
| 1994 | 0.25[b] **(0.31)** | 3.4 (4.2) | 0.4 (0.6) | | 0.08 (0.10) | 0.250[c] | | | | |
| 1995 | | | | | | 0.231 | | 0.4 | 2 (0.05) | |
| 1996 | | | | 2 (0.05) | | 0.225 | | 0.4 (0.6)[c] | | |
| 2003 | | | | | | 0.062 | ~ 2.4[d] | ~ 0.3[d] | | |
| | Tier 2 U.S. (2004–2006 interim; 2007 final) and LEV II California standards (2004 + ) | | | | | | | | | |
| 2004 | 0.125 (0.156) | | 0.3[e] | 0.95 (0.05) | 0.08 (0.08) | 0.053 | ~ 2.0 | ~ 0.07 | 0.5 (0.05) | ~ 0.01[d] |
| 2007 | < 0.09[e] | | 0.07 | | 0.02 (0.02) | 0.043 | ~ 1.8 | ~ 0.06 | | ~ 0.01 |
| 2010 | | | | | | 0.035 | ~ 1.7 | ~ 0.05 | | ~ 0.01 |

**Table B-1. U.S. Federal and California Light-duty Standards (g/mi units unless noted otherwise)**

*Note*: LEV = Low Emission Vehicle (Program)

[a] Evaporative units are grams/test (g/test) or, if in parentheses, grams/mile (g/mi). Tests changed in 1978. Beginning in 1995 in California (1996 federally), emissions could not exceed 2.0 g HC from a three-day stationary test, plus 0.05 g/mi while operating. Beginning in 2004, three-day standards became more stringent

[b] Standards for five years or 50,000 miles. Tier 1 is 10 years or 100,000 miles (in parentheses). Beginning in 1993, California required 100,000-mile standards for nonmethane HC and CO (in parentheses); some vehicles met 120,000-mile standards beginning in 1992 (not shown)

[c] From 1994 to 2010, LEV I and LEV II required declining nonmethane organic gas (HC) emissions; the table shows example years. A 100,000-mile $NO_x$ standard began in 1996

[d] The California CO, $NO_x$, and PM standards shown for 2003 and newer vehicles are approximated here by scaling from 120,000-mile nonmethane organic gas requirements

[e] Tier 2 required 0.30 g/mi $NO_x$ for 2004–2006 vehicles (0.07 g/mi starting in 2007). EPA anticipated that fleets meeting 0.07 g/mi $NO_x$ would emit less than 0.09 g/mi nonmethane organic gas. Standards were 10 years or 120,000 miles

*Source:* Calvert et al. 1993; U.S. EPA 1993a, 2000c, 2001b; Godish 1997; CARB 1999a, 1999b, 1999c, 2001; BTS 2001

encourage HC and CO production (Godish, 1997). Historically, manufacturers therefore were challenged to develop emission control strategies that reduced HC, CO, and $NO_x$ emissions at the same time and under various driving conditions. Similarly, I/M test regimes were challenged to adequately represent a range of real-world driving and emissions.

## NEW-VEHICLE EMISSION STANDARDS AND CONTROLS

Once the motor vehicle's contribution to air pollution was recognized, government agencies sought to reduce emissions. The key to these efforts focused on mandating new-vehicle emission standards. Table B-1 presents federal and California emission standards over time.

To implement exhaust standards, EPA created a certification test to measure emissions from newly manufactured cars. The test required vehicles to be driven on a dynamometer while the driver followed a prescribed course meant to simulate real-world driving. Known as the Federal Test Procedure (FTP), it first applied to the 1972 model year. The test cycle was based on home-to-work commuting patterns observed in Los Angeles during the mid-1960s. The FTP is similar in

| *Model years* | *Control stages or conditions* |
|---|---|
| 1968 | Federal emission control era begins (1966 in California). Emission standards must be met for a vehicle useful life of 5 years or 50,000 miles. |
| 1971 | Evaporative emission control canisters installed on U.S. vehicles. |
| 1972 | FTP-style emission testing begins to certify new vehicles. |
| 1973 | $NO_x$ controls added. |
| 1975 | Introduction of catalytic converters (oxidation catalysts) and lead-free fuel. U.S. evaporative emission standards begin. |
| 1978 | Evaporative emission certification testing begins, based on stationary vehicle tests. |
| 1981 | Three-way catalysts are widely introduced to meet more stringent CO and $NO_x$ standards. (Volvo introduced first three-way catalyst with the 1977 model year.) |
| 1984–1985 | Vehicles produced include some that are fuel injected (remainder are carbureted). |
| 1986 | All vehicles are fuel injected and improved emission control systems are introduced. |
| 1994 | Controls related to Tier 1 (U.S.) and LEV I (California) vehicle standards begin. Useful life standards extend to 10 years or 100,000 miles. |
| 1996 | Enhanced evaporative emission certification testing begins. All vehicles are equipped with second-generation on-board diagnostic (OBD) systems. |
| 2000 | Supplemental FTP (SFTP) certification testing added (2001 in California). |
| 2004 | Controls related to Tier 2 (U.S.) and LEV II (California) vehicle standards begin. Useful life standards set for 10 years or 120,000 miles. |

**Table B-2. Evolution of U.S. Light-duty Vehicle Emission Controls, by Model Year**
*Sources*: Cadle et al. 1998a, 1998b; Pierson et al. 1999; U.S. EPA 1996b; CARB 1999b, 2000a, 2000c; NRC 2001

concept to tests used by other nations (U.S. EPA 1995b; Degobert 1995; Chrysler Corporation 1998).

Over time, EPA and others recognized that the FTP failed to capture the full range of real-world driving behavior. California and U.S. regulators then implemented a Supplemental Federal Test Procedure (SFTP), which included more aggressive travel behavior (faster accelerations) and air conditioner operation. The SFTP requirements for new light-duty vehicles were phased in from model year 2000 through 2002; California's SFTP was phased in from model year 2001 through 2004 (U.S. EPA 1996b, 2002; CARB 1999b).

As emission standards became more stringent, manufacturer control strategies evolved. This evolution led to model years when certain controls were introduced or improved. The concept of grouping motor vehicles by model year is important for I/M, as there is substantial interest in exempting model-year groups less likely to contribute to overall emissions and identifying subsets of the fleet to test more frequently. Thus it is useful to categorize the vehicle fleet into groups based on their pollution control equipment and resulting emissions. Table B-2 provides a summary of emission control development stages. Note, however, that research has long shown that malfunctioning vehicles are found across all model years, and that a small fraction of high-emitting vehicles contributes a disproportionate amount of fleet emissions (e.g., Lawson et al. 1990; NRC 2000, 2004).

APPENDIX C

# AN OVERVIEW OF THE U.S. RSD EXPERIENCE

The use of remote sensing devices (RSD) following the California-EPA debate provides an interesting side story to the U.S. I/M policy experience. The technology's application did not unfold as its advocates hoped, yet it made important contributions to the understanding of automotive pollution. Many in the air quality community, including several California elected officials, had hoped that RSD could inexpensively augment or even replace traditional vehicle tests. Over time, Arizona, Colorado, Missouri, Texas, and Virginia implemented RSD programs to complement their tailpipe and OBD tests.[1] California, which led the nation by piloting the first use of RSD, did not implement a routine RSD program, though the state continued to pilot its use in different locations. This appendix takes a closer look at the experiences of those states that experimented with RSD.[2]

## HIGH-EMITTER AND CLEAN-SCREEN PROGRAMS

Arizona, Texas, and Virginia implemented high-emitter RSD programs. These programs sought to identify high-emitting vehicles and require them to pass an out-of-cycle test, meaning one independent of the routine inspection. In the 1990s and early 2000s, there was widespread recognition that 5 to 10 percent of the vehicle fleet was responsible for a disproportionate amount of emissions (NRC 2001). Despite the likely pool of problem cars, however, the Arizona, Texas, and Virginia RSD programs identified only a minute fraction of the on-road fleet as high-emitting: 0.15, 0.03, and 0.08 percent, respectively. Real-world experience showed that RSD implementation was handicapped by several difficulties, including having to use very high emission levels as a testing threshold, to avoid false failures, and taking emission measurements at locations that favored the use of later-model, cleaner-operating vehicles. The results proved so disappointing that Arizona, the first state to launch a comprehensive RSD program, abandoned the program after five years. Table C-1 summarizes the experiences of the states implementing RSD as part of their regular I/M programs.

In contrast to high-emitter programs, Missouri pioneered RSD use to exempt low emitters from routine inspections, a process called clean screening. Colorado and Virginia implemented clean-screen programs as well; however, these programs fared little better than their high-emitter counterparts. Although the goal of Colorado's RSD program was to exempt up to 80 percent of the fleet from routine inspections, it was able to exempt just 6 percent. Virginia capped the number of clean-screened vehicles not to exceed the high emitters identified; thus, given the few high emitters found, clean screening exempted a small number of vehicles. A bright spot was Missouri's successful use of RSD to clean-screen up to 29 percent of its fleet. However, clean-screen programs in Missouri and Virginia were careful to limit eligibility to model years 1995 and older, to avoid problems with OBD-equipped vehicles (VA DEQ 2006b). OBD problems arose because RSD clean screens exempted vehicles based solely on their tailpipe emissions. Clean screening prevented checking OBD data that identified excessive fuel evaporation, problematic cold-start conditions, and emerging problems that could later trigger excessive exhaust. Missouri abandoned its clean-screen program in late 2007 because of its incompatibility with OBD.

Even if OBD had not limited RSD's usefulness, it is unlikely that RSD would have seen widespread use as a clean-screen resource. Early research showed that simply exempting the newest model years from I/M was a far simpler and cost-effective method of clean screening (Wenzel and Sawyer 1998); later work confirmed that in systems already exempting new vehicles from I/M, clean screening was not cost-effective (Burnette et al. 2008).

| | *Arizona* | *Missouri* | *Colorado* | *Texas* | *Virginia* |
|---|---|---|---|---|---|
| Background | Arizona implemented the first comprehensive U.S. program (1995–2000). In 1999, costs were $1 million per year and more than $300 per high emitter identified. | Missouri started the first U.S. clean-screen program, Rapid Screen, in 1999. It reduced the number of vehicles subject to routine inspections. | In 2004, the Denver area began a Rapid Screen program similar to Missouri's. The goal was to screen up to 80% of vehicles from I/M. | In 1999, Texas began RSD use to identify high emitters. As of 2007, it deployed seven units, rotated among 450 sites. The program identified about 200 vehicles per month. | In 2002, Virginia began using RSD for data analysis, and in 2006, it began high-emitter and clean-screen programs. Approximate costs were $1 million for setup, $700,000 for annual operations. |
| Assessments and problems | False failure and measurement problems. From 1998 to 1999, it identified only 3,000 high emitters from more than 2 million vehicles and produced few emission reductions. | As of 2002, RSD exempted from I/M 25%–29% of St. Louis vehicles. In 2005, officials recommended ending Rapid Screen, based partly on its OBD incompatibility. | In 2005, of 900,000 vehicles subject to I/M, Rapid Screen identified only 3% as eligible for an I/M exemption. | Initially, a legislator opposed RSD over privacy concerns, and one official declined to prosecute high-emitter cases. State officials eventually overcame these difficulties. | Clean screens were capped at the number of high emitters; they were also limited to 1995 and older vehicles to prevent OBD incompatibility. |
| Outcomes | The state repealed the program by 2000. In 2006, it found, despite improved technology, that a regulatory program remained premature. | Rapid Screen ended in 2007. | By 2006, an expanded Rapid Screen screened about 6% of vehicles from required testing. | As of 2007, RSD identified only 0.03% of the fleet as high emitting because of stringent failure thresholds and cleaner vehicles operating near RSD sites. | In its first months, RSD found 43 high emitters out of 53,379 vehicles (0.08%), less than the 2% expected. Units were mostly deployed where vehicles tended to be newer. |

**Table C-1. U.S. RSD Implementation Experiences in Five States**

*Source:*

Arizona: Arizona State Legislature 1999; ADEQ 2000; NRC 2001; Eastern Research Group 2002, 2006
Missouri: Klausmeier 2002; MO DNR 2005; VA DEQ 2006b; State of Missouri 2007
Colorado: RAQC 2000; Pollack et al. 2003; Klausmeier et al. 2006
Texas: Guckian 2007
Virginia: Klausmeier and McClintock 2003; McClintock 2006; VA DEQ 2006b, 2007

There is perhaps no better evidence about the unfulfilled promise of RSD than the history of the firms engaged in marketing its use. A California study found that although several companies formed to sell RSD technologies in the mid-1990s, as of 2008 only one firm, Environmental System Products, remained a major presence in the U.S. RSD market (Burnette et al. 2008).

## OTHER CONTRIBUTIONS

Despite its failure to contribute meaningfully to high-emitter identification in I/M programs and its limitations as a clean-screen resource, RSD use offered unique insights about the vehicle fleet as a whole. The National Research Council used RSD-based studies to draw comparisons between real-world I/M effects and EPA model predictions. Virginia used RSD to compare vehicle emissions from cars operating in and outside of I/M program areas. RSD-based analyses identified that beginning with 1996 vehicles, fleet emissions dropped as a result of the introduction of OBD technology (NRC 2001; Klausmeier and McClintock 2003; Bishop et al. 2006).

RSD also helped document that, contrary to computer predictions, as a particular group of vehicles aged—for example, all model year 1975 vehicles—at some point, perhaps after about 15 years, average emissions from that model-year group decreased rather than increased. It showed that the last vehicles left from a particular model year were typically the ones that were best maintained; higher emitters, likely because of poor maintenance, left the fleet earlier (Stedman and Bishop 2004).

In Denver, RSD readings educated the driving public about their vehicle emissions. After vehicles passed an RSD unit that measured their emissions, they passed a roadside electronic sign where emission results were posted in real time (Bishop et al. 2000). Perhaps the most lasting contribution made by RSD use was its ability to consistently demonstrate, over many years and in many countries, that for a given pollutant, a small fraction of vehicles contributed a disproportionate amount of emissions (e.g., Lawson et al. 1990; Sjodin and Lenner 1995; Schifter et al. 2003).

## CALIFORNIA EXPERIMENTATION WITH RSD

Although the 1994 California-EPA deal triggered the first large-scale use of RSD, in the Sacramento pilot project, the state did not implement an ongoing program. In 2000, when state assessments found that Smog Check fell short of its emission reduction goals, California committed to implement another RSD pilot to test clean screening and high-emitter identification (Kenny 2000). During 2004 and 2005, the second pilot study collected more than 2 million RSD measurements. Researchers reconfirmed RSD's limitations and forecast the following results if California chose to implement a more widespread program (Burnette et al. 2008):

- RSD would measure only 30 percent of the I/M fleet, even assuming deployment of 50 units (a program scale that would constitute the nation's most comprehensive use of RSD); for comparison, programs in Colorado, Missouri, Texas, and Virginia deployed from two (Virginia) to eight (Colorado) units (McClintock and Vescio 2007).
- Less than 50 percent of vehicles identified by RSD as high-emitting would fail a California I/M exhaust emission test. The differences would be due to repairs obtained between the RSD reading and the Smog Check, variability in Smog Check test accuracy, and other factors.
- The cost-effectiveness of a high-emitter program would be poor: from about $42,000 to $80,000 per ton of pollutant reduced. Clean screening also would not be cost-effective.

In February 2005, air quality regulators in the Los Angeles area dedicated $4 million to implement a short-term RSD program in the worst-polluted U.S. region. They planned to deploy RSD units for one year, develop a database of high-emitting vehicles, and use financial incentives (limited to $1,000 per vehicle) to encourage up to 6,000 owners to repair or scrap their vehicles (SCAQMD 2005, 2006; Saito 2007). Findings as of mid-2007 indicated that of the vehicles measured by RSD, 15 percent of those from model years 1995 and older, and 1 percent of those from 1996 and newer had been identified as high-emitting. The findings were further confirmation that a relatively small fraction of vehicles contributed overwhelmingly to overall fleet emissions (McClintock 2007).

## NOTES

1. Georgia launched an RSD program in 1996 that collected data but did not affect vehicle inspections.
2. Some of the findings presented here were originally published in Eisinger and Wathern 2008; they are reprinted with permission from Elsevier.

# INDEX

*Note: Page numbers in italics indicate figures and tables. Page numbers followed by an 'n' indicate notes.*

# ABOUT THE AUTHOR

Douglas S. Eisinger is the director of transportation policy and planning at Sonoma Technology, Inc., an air quality research firm. He has served for more than 12 years as the program manager for the U.C. Davis–Caltrans Air Quality Project; he is also an adjunct associate professor at the University of Hawaii. From 1991 to 1995, he was mobile sources section chief for the U.S. Environmental Protection Agency in San Francisco. From 2006 to 2007, he was a fellow in environmental regulatory implementation at Resources for the Future.

Since the 1980s, Eisinger has completed numerous transportation-related air quality studies, including projects sponsored by EPA, the U.S. Federal Highway Administration, California Air Resources Board, California Department of Transportation, California Bureau of Automotive Repair, South Coast Air Quality Management District, and other government agencies. He is a member of the U.S. Transportation Research Board, Transportation and Air Quality Committee. In 2006, he was selected as a national transportation–air quality expert by the Center for Environmental Excellence at the American Association of State Highway and Transportation Officials.

Eisinger has a Ph.D. in environmental policy analysis from the University of Wales in the United Kingdom, a Masters in public policy from Harvard University's John F. Kennedy School of Government, and a Bachelor's degree in government from Cornell University.